On Necessary and Sufficient Conditions for L^p-Estimates of Riesz Transforms Associated to Elliptic Operators on $\mathbb{R}^n$ and Related Estimates

Memoirs
of the
American Mathematical Society

Number 871

On Necessary and Sufficient Conditions for L^p-Estimates of Riesz Transforms Associated to Elliptic Operators on $\mathbb{R}^n$ and Related Estimates

Pascal Auscher

March 2007 • Volume 186 • Number 871 (first of five numbers) • ISSN 0065-9266

American Mathematical Society
Providence, Rhode Island

2000 *Mathematics Subject Classification.* Primary 42B20, 42B25, 47F05, 47B44, 35J15, 35J30, 35J45.

Library of Congress Cataloging-in-Publication Data

Auscher, Pascal,
    On necessary and sufficient conditions for L^p-estimates of Riesz transforms associated to elliptic operators on $\mathbb{R}^n$ and related estimates / Pascal Auscher.
        p. cm. — (Memoirs of the American Mathematical Society, ISSN 0065-9266 ; no. 871)
    "Volume 186, number 871 (first of 5 numbers)."
    Includes bibliographical references.
    ISBN-13: 978-0-8218-3941-6 (alk. paper)
    ISBN-10: 0-8218-3941-1 (alk. paper)
        1. Singular integrals.   2. Littlewood-Paley theory.   3. Calderón-Zygmund operator.   4. Elliptic operators.   5. Semigroups.   I. Tilte.

QA403.5.A87   2007
515'.23—dc22
                                                         2006047929

Memoirs of the American Mathematical Society

This journal is devoted entirely to research in pure and applied mathematics.

Subscription information. The 2007 subscription begins with volume 185 and consists of six mailings, each containing one or more numbers. Subscription prices for 2007 are US$649 list, US$519 institutional member. A late charge of 10% of the subscription price will be imposed on orders received from nonmembers after January 1 of the subscription year. Subscribers outside the United States and India must pay a postage surcharge of US$38; subscribers in India must pay a postage surcharge of US$43. Expedited delivery to destinations in North America US$53; elsewhere US$130. Each number may be ordered separately; *please specify number* when ordering an individual number. For prices and titles of recently released numbers, see the New Publications sections of the *Notices of the American Mathematical Society*.

Back number information. For back issues see the *AMS Catalog of Publications*.

Subscriptions and orders should be addressed to the American Mathematical Society, P. O. Box 845904, Boston, MA 02284-5904, USA. *All orders must be accompanied by payment.* Other correspondence should be addressed to 201 Charles Street, Providence, RI 02904-2294, USA.

Copying and reprinting. Individual readers of this publication, and nonprofit libraries acting for them, are permitted to make fair use of the material, such as to copy a chapter for use in teaching or research. Permission is granted to quote brief passages from this publication in reviews, provided the customary acknowledgment of the source is given.

Republication, systematic copying, or multiple reproduction of any material in this publication is permitted only under license from the American Mathematical Society. Requests for such permission should be addressed to the Acquisitions Department, American Mathematical Society, 201 Charles Street, Providence, Rhode Island 02904-2294, USA. Requests can also be made by e-mail to reprint-permission@ams.org.

Memoirs of the American Mathematical Society is published bimonthly (each volume consisting usually of more than one number) by the American Mathematical Society at 201 Charles Street, Providence, RI 02904-2294, USA. Periodicals postage paid at Providence, RI. Postmaster: Send address changes to Memoirs, American Mathematical Society, 201 Charles Street, Providence, RI 02904-2294, USA.

© 2007 by the American Mathematical Society. All rights reserved.
This publication is indexed in *Science Citation Index*®, *SciSearch*®, *Research Alert*®, *CompuMath Citation Index*®, *Current Contents*®/*Physical, Chemical & Earth Sciences*.
Printed in the United States of America.

∞ The paper used in this book is acid-free and falls within the guidelines established to ensure permanence and durability.
Visit the AMS home page at http://www.ams.org/

10 9 8 7 6 5 4 3 2 1   12 11 10 09 08 07

Contents

Acknowledgements	ix
Introduction	xi
Notation	xvii

Chapter 1. Beyond Calderón-Zygmund operators	1

Chapter 2. Basic L^2 theory for elliptic operators	9
2.1. Definition	9
2.2. Holomorphic functional calculus on L^2	9
2.3. L^2 off-diagonal estimates	10
2.4. Square root	12
2.5. The conservation property	12

Chapter 3. L^p theory for the semigroup	15
3.1. Hypercontractivity and uniform boundedness	15
3.2. $W^{1,p}$ elliptic estimates and hypercontractivity	17
3.3. Gradient estimates	19
3.4. Summary	21
3.5. Sharpness issues	22
3.6. Analytic extension	22

Chapter 4. L^p theory for square roots	25
4.1. Riesz transforms on L^p	25
4.2. Reverse inequalities	32
4.3. Invertibility	36
4.4. Applications	37
4.5. Riesz transforms and Hodge decomposition	38

Chapter 5. Riesz transforms and functional calculi	41
5.1. Blunck & Kunstmann's theorem	41
5.2. Hardy-Littlewood-Sobolev estimates	42
5.3. The Hardy-Littlewood-Sobolev-Kato diagram	44
5.4. More on the Kato diagram	47

Chapter 6. Square function estimates	51
6.1. Necessary and sufficient conditions for boundedness of vertical square functions	51
6.2. On inequalities of Stein and Fefferman for non-tangential square functions	60

Chapter 7. Miscellani		65
7.1. Local theory		65
7.2. Higher order operators and systems		66
Appendix A. Calderón-Zygmund decomposition for Sobolev functions		69
Appendix. Bibliography		73

Abstract

This memoir focuses on L^p estimates for objects associated to elliptic operators in divergence form: its semigroup, the gradient of the semigroup, functional calculus, square functions and Riesz transforms. We introduce four critical numbers associated to the semigroup and its gradient that completely rule the ranges of exponents for the L^p estimates. It appears that the case $p < 2$ already treated earlier is radically different from the case $p > 2$ which is new. We thus recover in a unified and coherent way many L^p estimates and give further applications. The key tools from harmonic analysis are two criteria for L^p boundedness, one for $p < 2$ and the other for $p > 2$ but in ranges different from the usual intervals $(1,2)$ and $(2,\infty)$.

Received by the editor January 6, 2004, Revised May 30, 2005.
2000 *Mathematics Subject Classification.* Primary 42B20, 42B25, 47F05, 47B44, 35J15, 35J30, 35J45.
Key words and phrases. elliptic operators, divergence form, semigroup, L^p estimates, Calderón-Zygmund theory, good lambda inequalities, hypercontractivity, Riesz transforms, holomorphic functional calculus, Littlewood-Paley-Stein estimates.

Acknowledgements

It is a pleasure to thank here many colleagues with whom I discussed these topics over the years and from whom I learnt a lot. Let me mention S. Blunck, T. Coulhon, G. David, X.T. Duong, S. Hofmann, A. McIntosh, E. Ouhabaz, P. Tchamitchian. Special thanks go to E. Ouhabaz who helped me with some historical comments and to S. Blunck for providing me with recent unpublished work. I also indebted to J.-M. Martell for a careful reading that helped correcting some misprints in the submitted version.

Introduction

Although the matter of this text applies *in extenso* to elliptic operators or systems in divergence form to any order in $\mathbb{R}^n$, we focus on second order operators in $\mathbb{R}^n$. A section will be devoted to these more general classes.

Let $A = A(x)$ be an $n \times n$ matrix of complex, L^∞ coefficients, defined on $\mathbb{R}^n$, and satisfying the ellipticity (or "accretivity") condition

$$\lambda |\xi|^2 \leq \operatorname{Re} A\xi \cdot \bar{\xi} \text{ and } |A\xi \cdot \bar{\zeta}| \leq \Lambda |\xi||\zeta|,$$

for $\xi, \zeta \in \mathbb{C}^n$ and for some λ, Λ such that $0 < \lambda \leq \Lambda < \infty$. We define a second order divergence form operator

$$Lf \equiv -\operatorname{div}(A\nabla f),$$

which we interpret in the sense of maximal accretive operators via a sesquilinear form. Here, ∇ denotes the array of first order partial derivatives.

The maximal accretivity condition implies the existence of an analytic contraction semigroup on $L^2(\mathbb{R}^n)$ generated by $-L$. It also implies the existence of a holomorphic functional calculus that has the expected stability under commutation and convergence, allowing for example to define fractional powers. This in turn yields the possibility of defining various objects of interest both from functional and harmonic analysis points of view. Let us mention Littlewood-Paley-Stein type functionals such as

$$g_L(f)(x) = \left(\int_0^\infty |(L^{1/2} e^{-tL} f)(x)|^2 \, dt \right)^{1/2}$$

and

$$G_L(f)(x) = \left(\int_0^\infty |(\nabla e^{-tL} f)(x)|^2 \, dt \right)^{1/2}.$$

The "singular integral" pending to the latter square function is the so-called Riesz transform associated to L given for example by

$$\nabla L^{-1/2} = \frac{1}{\sqrt{\pi}} \int_0^\infty \sqrt{t} \, \nabla e^{-tL} \, \frac{dt}{t}.$$

Other objects of interest are 1) the operator of maximal regularity for the parabolic equation associated to L, 2) Riesz means and L^p-multipliers when L is, in addition, self-adjoint ... They can be treated by the methods presented here but we have chosen not to go into such developments.

When the coefficients are constant, *e.g.* the Laplacian, one finds classical objects in harmonic analysis: multipliers, the Littlewood-Paley-Stein functionals and the original Riesz transforms. They belong to the well-understood class of Calderón-Zygmund operators. If the coefficients of L still have some smoothness, then the tools of pseudo-differential calculus or of Calderón-Zygmund theory can still be used. In absence of regularity of the coefficients, these operators fall beyond the above classes and this participate to Calderón's program [1] of defining algebras of differential operators with minimal smoothness.

The first step of that study is the action on $L^2(\mathbb{R}^n)$. First, there is a bounded holomorphic functional calculus on L^2 basically as a consequence maximal accretivity and Von Neumann's inequality. One has

$$\|\varphi(L)f\|_2 \leq \|\varphi\|_\infty \|f\|_2$$

for φ bounded holomorphic in the open right half plane. Next, g_L and G_L are L^2 bounded (see Chapter 6) and

$$\|g_L(f)\|_2 \sim \|f\|_2 \sim \|G_L(f)\|_2.\ [2]$$

The L^2 boundedness of the Riesz transform has been proved recently and in fact, one has in all dimensions

$$\|L^{1/2}f\|_2 \sim \|\nabla f\|_2.$$

This implies that the domain of $L^{1/2}$ is the Sobolev space $W^{1,2}$, which was known as Kato's conjecture. [3]

The second step is the action on $L^p(\mathbb{R}^n)$ with $1 < p < \infty$ but $p \neq 2$. The bounded holomorphic functional calculus on L^p consists in proving L^p boundedness of $\varphi(L)$ for an appropriate class of bounded holomorphic functions φ. It is completely understood [4] and one has

$$\|\varphi(L)(f)\|_p \lesssim \|\varphi\|_\infty \|f\|_p \quad \text{whenever } p_-(L) < p < p_+(L),\ [5]$$

where $p_-(L)$ and $p_+(L)$ are the two critical exponents for the L^p uniform boundedness of the semigroup $(e^{-tL})_{t>0}$. It is clear that this interval is the largest open range of exponents for which such an inequality holds as φ can be an exponential. The L^p theory for square functions consists in comparing the L^p norms of $g_L(f)$, f and $G_L(f)$. For g_L, what happens is completely understood in terms of functional calculus: [6] one has

$$\|g_L(f)\|_p \sim \|f\|_p \quad \text{whenever } p_-(L) < p < p_+(L).$$

It turns out that this interval is the largest open range of exponents for which this equivalence holds. The comparison between the L^p norms f and $G_L(f)$ has not been done in general so far [7] and we shall see that

$$\|G_L(f)\|_p \sim \|f\|_p \quad \text{whenever } q_-(L) < p < q_+(L),$$

---

[1] See [23].

[2] Here $\sim$ is the equivalence in the sense of norms, with implicit constants C depending only on n, λ and Λ.

[3] The one dimensional Kato's conjecture (that is the case $n = 1$)) is first proved by Coifman, McIntosh & Meyer [25, Théorème X] the two dimensional case by Hofmann & McIntosh [42, Theorem 1.4] and the general case in any dimension by Hofmann, Lacey, McIntosh & Tchamitchian along with the author [10, Theorem 1.4]. We refer to the latter reference for historical remarks and connections with other problems.

[4] This is essentially due to Blunck & Kunstmann [21, Proposition 2.3] although the authors did not introduce the numbers $p_\pm(L)$.

[5] Here $\lesssim$ is the comparison in the sense of norms, with implicit constant C that may depend on L through ellipticity, dimension, its type and $p_\pm(L)$.

[6] This follows by combining works of Blunck & Kunstmann [21, Proposition 2.3] and Le Merdy [48, Theorem 3].

[7] After this paper was submitted, the author learned of a work by Yan [68] where the inequality $\lesssim$ is obtained for $2n/(n+2) < p \leq 2$.

where $q_-(L)$ and $q_+(L)$ are the two critical exponents for the L^p uniform boundedness of the gradient of the semigroup $(\sqrt{t}\,\nabla e^{-tL})_{t>0}$, and that this open range is optimal. We also study the corresponding non-tangential Littlewood-Paley-Stein functionals (See Chapter 7).

The L^p theory for square roots consists in comparing $L^{1/2}f$ and ∇f in L^p norms. [8] There are two issues here, namely the Riesz transform L^p boundedness, that is an inequality $\|\nabla f\|_p \lesssim \|L^{1/2}f\|_p$, and its reverse $\|L^{1/2}f\|_p \lesssim \|\nabla f\|_p$. It turns out the ranges of p may be different. The state of the art for this class of operators L is as follows. [9] One has the Riesz transforms estimates [10]

$$\|\nabla f\|_p \lesssim \|L^{1/2}f\|_p \quad \text{if} \quad \begin{cases} n=1 & \text{and} \quad 1<p<\infty \\ n=2 & \text{and} \quad 1<p<2+\varepsilon \\ n\geq 3 & \text{and} \quad \frac{2n}{n+2}-\varepsilon<p<2+\varepsilon \end{cases},$$

and the reverse inequalities [11]

$$\|L^{1/2}f\|_p \lesssim \|\nabla f\|_p \quad \text{if} \quad \begin{cases} n=1,2 & \text{and} \quad 1<p<\infty \\ n=3,4 & \text{and} \quad 1<p<\frac{2n}{n-2}+\varepsilon \\ n\geq 5 & \text{and} \quad \frac{2n}{n+4}-\varepsilon<p<\frac{2n}{n-2}+\varepsilon \end{cases}.$$

Of course, if L possesses more properties then the ranges of exponents p improve. For example for constant coefficients operators these inequalities holds when $1<p<\infty$, and for real operators in dimensions $n\geq 3$, the Riesz transform L^p boundedness is valid for $1<p<2+\varepsilon$ and the reverse inequality for $1<p<\infty$. [12] Hence, it is interesting to have a theory that works for any single operator. In fact, the conclusion of the story for the Riesz transform is [13]

$$\|\nabla f\|_p \lesssim \|L^{1/2}f\|_p \quad \text{if and only if} \quad q_-(L)<p<q_+(L)$$

---

[8] This program was initialised by the author and P. Tchamitchian in [17] for this class of complex operators. It arose from a different perspective towards applications to boundary value problems in the works of Dalbergh, Jerison, Kenig and their collaborators (see [47, problem 3.3.16]).

[9] Some of the results were obtained prior to the Kato conjecture by making the L^2 result an assumption.

[10] For $n = 1$, this is due to the author and Tchamitchian [18, Théorème A] for $n = 2$ to the author and Tchamitchian [17, Chapter IV, Theorem 1] combining the Gaussian estimates of the author, McIntosh and Tchamitchian [15, Theorem 3.5] and the L^2 result [10, Theorem 1.4] and for $n \geq 3$ and $p_n = \frac{2n}{n+2} < p < 2$, independently to Blunck & Kunstmann [21, Theorem 1.2], and to Hofmann & Martell [42, Theorem 1.2]. The enlargement of the range below p_n is due to the author [6, Proposition 1] and above 2 is a consequence of the method of the author and Tchamitchian once the reverse inequality is established (see [17, Chapter IV, Proposition 20] and [6, Corollary 4]).

[11] They are due, for $n = 1$, to the author and Tchamitchian ([18], Théorème A), for $n = 2$, to the author and Tchamitchian ([17], Chapter IV, Theorem 1), again combining [15], Theorem 3.5 with [10], Theorem 1.4, for $n = 3, 4$ to the author, Hofmann, Lacey, McIntosh & Tchamitchian ([10], Proposition 6.2) and, for $n \geq 5$ to the author ([6], Theorem 2).

[12] For constant coefficients, this goes back to Calderón-Zygmund original work [24] and for real operators, this is due to the author and Tchamitchian [17, Chapter IV].

[13] This problematic of finding the "smallest" exponent p is implicit in [21] and we present the counterpart for the "largest", which turns out to require different arguments. After this paper was submitted, Shen [59] informed me of his independent and simultaneous work on the same problem for $p > 2$ when L is real. He obtains a different characterization of $q_+(L)$ in terms of reverse Hölder estimates for weak solutions of $Lu = 0$. We convince ourselves by e-mail discussions that this approach can be adapted to complex L.

and we also show that

$$\|L^{1/2}f\|_p \lesssim \|\nabla f\|_p \quad \text{whenever } \tilde{p}_-(L) = \sup(1, (p_-(L))_*) < p < p_+(L).$$

This encapsulates all the above mentioned estimates (See Notation for the meaning of p_*). Concerning the latter range, we show that $p_+(L)$ is best possible in some sense, while we only know a lower bound on $\tilde{p}_-(L)$. [14] Staring at the formula given above for computing the Riesz transform this result seems to say that the integral yields a bounded operator on L^p if and only if the integrands are uniformly bounded operators on L^p. Said like this, the sufficiency looks astonishingly simple. But this is not quite the truth as there is a play on exponents in the proof. Note also the range of exponents for L^p-boundedness of the Riesz transform $R(L) = \nabla L^{-1/2}$ is characterized. In particular, it is an open set. We also show that $2 < p < q_+(L)$ if and only if the Hodge projector $\nabla L^{-1}\,\mathrm{div}$ (or alternately, the second order Riesz transform $R(L)R(L^*)^*$) is bounded on L^p. For $p < 2$ the L^p boundedness of the Hodge projector is stronger than that of $R(L)$.

The objective of this paper is to present a complete, coherent and unified theory for all these topics. We present works of others and also original contributions. In particular, we have tried to be self-contained. Our main observation is the following: **four critical numbers** [15] **rule the L^p behavior.** These are $p_\pm(L)$, the limits of the interval of exponents $p \in [1, \infty]$ for which the semigroup $(e^{-tL})_{t>0}$ is L^p bounded, and $q_\pm(L)$, the limits of the interval of exponents $p \in [1, \infty]$ for which the family $(\sqrt{t}\,\nabla e^{-tL})_{t>0}$ is L^p bounded. We make a thorough study of these numbers, their inner relationships and their values in terms of dimension for the whole class of such L (Chapter 4). The key stone of this theory, which makes use of the divergence structure of our operators, is that the L^p boundedness of their semigroups (resp. of the gradient of their semigroups) is **equivalent** to some **off-diagonal estimates** and also to some **hypercontractivity**. [16] To be precise, we sometimes have to loosen the exponent p, but this is harmless for the kind of results we are after. Let us mention here that this equivalence is not powerful enough for treating L^∞ decay of the semigroup kernels whenever there is L^1 or L^∞ boundedness of the semigroup. But, again, this is enough for our needs.

Next, we turn to studying the harmonic analysis objects (Chapters 5, 6 & 7). On the one hand, finding necessary conditions on p for which one has L^p bounds for the functional calculus, the square functions, the Riesz transforms is intuitively easy and the critical numbers appear then. On the other hand, it is not clear at all why these conditions **alone** suffice. For this, appropriate criteria for L^p boundedness with minimal hypotheses are needed.

The L^p estimates obtained in Chapters 5, 6 & 7 depend on the critical numbers of a given operator. Thus, they are individual results with sharp ranges of p's, the operator norms depending on dimension, the ellipticity constants and some of the critical numbers. But when the critical numbers can be estimated for operators

---

[14] As said above, all applies to higher-order operators. For that extended class, this lower bound is optimal due to some existing appropriate counter-examples in the theory. Similar counter-examples for the second order case are not known.

[15] In fact they reduce to two: $p_-(L)$ and $q_+(L)$.

[16] This equivalence for values of p different from 1, although not explicitely stated in the literature to our knowledge, is in the air of a number of works dealing with semigroups of elliptic operators. It appears first in Davies's work [29]. See also [50] and the references therein.

in some class, they become L^p estimates for the whole class. In this case, the optimality of the range of p's is function of the sharpness of the bounds on the critical numbers. This is discussed in Section 3.5 for second order operators and in Section 7.2 for higher order.

As the reader may guess, the various critical numbers have no reason to be 1 or ∞. Hence, we have a class of operators that lies **beyond the class of Calderón-Zygmund operators**. We wish here to present an appropriate machinery to obtain L^p boundedness without caring about kernels of operators and for ranges of p different from the usual intervals $(1, 2)$ or $(2, \infty)$ (See Chapter 1 for more).

Notation

We constantly work on $\mathbb{R}^n$, $n \geq 1$, equipped with its usual Lebesgue measure. If E is a measurable set in $\mathbb{R}^n$, we write

$$\|f\|_{L^p(E)} = \left(\int_E |f|^p\right)^{1/p}$$

for the norm in the Lebesgue space $L^p(E)$, $1 \leq p \leq \infty$, with the usual modification if $p = \infty$. We do not indicate the integration variable and the measure unless this is necessary for comprehension. Also we drop E in the lower limit if $E = \mathbb{R}^n$ and set L^p and $\|f\|_p$ for $L^p(\mathbb{R}^n)$ and $\|f\|_{L^p(\mathbb{R}^n)}$ unless the value of n is of matter. For Hilbert-valued functions, $|f|$ is replaced by the norm in the Hilbert space, $|f|_H$, but we do not introduce a specific notation for L^p as the situation will make it clear.

We use the notation p' for the dual exponent to p: $p' = \frac{p}{p-1}$.

The Sobolev space $W^{m,p}(\mathbb{R}^n)$, $m \in \mathbb{N}$, $1 \leq p \leq \infty$, is the space of those L^p functions f for which all derivatives up to and including order m are in L^p. The norm of f is the sum of the L^p norms of f and all its derivatives.

The homogeneous Sobolev space $\dot{W}^{m,p}(\mathbb{R}^n)$, $m \in \mathbb{N}$, $1 \leq p < \infty$, is the closure of $C_0^\infty(\mathbb{R}^n)$ for the seminorm being the sum of the L^p norms of all derivatives of order m.

We are essentially interested in Sobolev spaces of order $m = 1$. The well-known Sobolev inequalities say that

$$\dot{W}^{1,p}(\mathbb{R}^n) \subset L^q(\mathbb{R}^n)$$

whenever, $n > 1$,

$$1 \leq p < n, \qquad \frac{n}{n-1} \leq q < \infty \qquad \frac{n}{q} = \frac{n}{p} - 1$$

and

$$\|f\|_q \leq C(n,p,q)\|\nabla f\|_p.$$

We use the notation p^* (p upper star) for the Sobolev exponent of p, that is

$$p^* = \frac{np}{n-p}$$

with the convention that $p^* = \infty$ if $p \geq n$ and q_* (q lower star) for the reverse Sobolev exponent of q, that is

$$q_* = \frac{nq}{n+q}.$$

Note that $(p_*)' = (p')^*$ whenever $p_* \geq 1$ and $(p^*)' = (p')_*$ whenever $p^* < \infty$.

For $1 \leq p, q \leq \infty$ we set

$$\gamma_{pq} = \left|\frac{n}{q} - \frac{n}{p}\right|$$

and in the special case where $q = 2$

$$\gamma_p = \left|\frac{n}{2} - \frac{n}{p}\right|.$$

As usual, we use positive constants which do not depend on the parameters at stake and whose value change at each occurence. Often, we do not mention about such constants as their meaning is self-explanatory.

CHAPTER 1

Beyond Calderón-Zygmund operators

By definition, [1] a Calderón-Zygmund operator on $\mathbb{R}^n$ [2] is a bounded operator on L^2 which is associated with a kernel possessing some size and regularity properties, the latter being called Hörmander's condition. The fundamental result [3] is that such operators are weak type $(1,1)$, hence strong type (p,p) when $1 < p < 2$ by the interpolation theorem of Marcinkiewicz and eventually strong type (p,p) for $p > 2$ by a duality argument. Another route is to begin with their $L^\infty - BMO$ boundedness, [4] interpolation between L^2 and BMO [5] for $2 < p < \infty$, and duality for $1 < p < 2$. However, one should not forget that this interpolation is more involved than the Calderón-Zygmund décomposition used for weak type $(1,1)$.

In applications, this is enough for numerous operators going from convolution operators such as the Hilbert transform and the classical Riesz transforms (the prototypes of Calderón-Zygmund operators) to the Cauchy integral on a Lipschitz curve and the double layer operator on a Lipschitz domain.

However, recently some interesting operators were found to be out of this class. That is they are strong type $(2,2)$ but the other properties fail. Some reasons are

(1) their kernel does not possess regularity properties such as the Hörmander condition
(2) they do not possess a kernel in any reasonable sense (but the distribution sense)
(3) they are found to be strong type (p,p) for a range of values of p different from $1 < p < 2$ or $2 < p < \infty$ or their unions
(4) duality does not apply

It is natural to ask the following question: **is there a general machinery to handle the L^p theory of such operators?**

The answer is yes. It turns out that the cases $p < 2$ and $p > 2$ are treated by different methods, which is useful when duality is not available.

Let us come now to statements. For simplicity, we work in the framework of $\mathbb{R}^n$ equipped with the Lebesgue measure, although the original results are presented in spaces of homogeneous types.

We denote as above by $B(x,r)$ the open (Euclidean) ball of radius $r > 0$ and center $x \in \mathbb{R}^n$, and set $|E|$ the measure of a set E.

---

[1] We take this terminology from Meyer [55, Chapter VII].

[2] or more generally on a space of homogeneous type (see [26]); we shall not be concerned with the development on non homogeneous spaces described in the excellent review by Verdera [66] although extensions to this setting of the results presented here would be interesting.

[3] This is due to Calderón & Zygmund [24] and Hörmander [43, Theorem 2.1] in the convolution case and is extended to non convolution operators in [26].

[4] This is attributed to Peetre-Spanne-Stein (see [63], p. 191)

[5] This result is due to Stampacchia [61].

Here is some further notation used throughout the paper. For a ball B, we let λB be the ball with same center and radius λ times that of B. We set
$$C_1(B) = 4B \quad \text{and} \quad C_j(B) = 2^{j+1}B \setminus 2^j B, \text{ if } j \geq 2.$$
We adopt the similar notation, λQ and $C_j(Q)$, for any cube Q with sides parallel to the axes.

Denote by M the Hardy-Littlewood maximal operator
$$Mf(x) = \sup_{B \ni x} \frac{1}{|B|} \int_B |f|,$$
where B ranges over all open balls (or cubes) containing x.

THEOREM 1.1. [6] *Let $p_0 \in [1, 2)$. Suppose that T is sublinear operator of strong type $(2, 2)$, and let A_r, $r > 0$, be a family of linear operators acting on L^2. Assume for $j \geq 2$*

$$(1.1) \quad \left(\frac{1}{|2^{j+1}B|} \int_{C_j(B)} |T(I - A_{r(B)})f|^2 \right)^{1/2} \leq g(j) \left(\frac{1}{|B|} \int_B |f|^{p_0} \right)^{1/p_0}$$

and for $j \geq 1$

$$(1.2) \quad \left(\frac{1}{|2^{j+1}B|} \int_{C_j(B)} |A_{r(B)}f|^2 \right)^{1/2} \leq g(j) \left(\frac{1}{|B|} \int_B |f|^{p_0} \right)^{1/p_0}$$

for all ball B with $r(B)$ the radius of B and all f supported in B. If $\Sigma = \sum g(j)2^{nj} < \infty$, then T is of weak type (p_0, p_0), with a bound depending only on the strong type $(2, 2)$ bound of T, p_0 and Σ, hence bounded on L^p for $p_0 < p < 2$.

THEOREM 1.2. [7] *Let $p_0 \in (2, \infty]$. Suppose that T is sublinear operator acting on L^2, and let A_r, $r > 0$, be a family of linear operators acting on L^2. Assume*

$$(1.3) \quad \left(\frac{1}{|B|} \int_B |T(I - A_{r(B)})f|^2 \right)^{1/2} \leq C \left(M(|f|^2) \right)^{1/2}(y),$$

and

$$(1.4) \quad \left(\frac{1}{|B|} \int_B |TA_{r(B)}f|^{p_0} \right)^{1/p_0} \leq C \left(M(|Tf|^2) \right)^{1/2}(y),$$

---

[6] This is due to Blunck & Kunstmann [20, Theorem 1.1] generalizing earlier work of Duong & McIntosh [33, Theorem 1] who obtained weak type $(1, 1)$ under a weakened Hörmander condition, still assuming reasonable pointwise estimates for kernels but no regularity in the classical sense. The statement and proof here simplify that of [20]. Note that when $p_0 = 1$ the assumptions are slightly different than those in [33]. One can find in Fefferman [36] weak type results for values of p not equal to one but no general statement is made.

[7] This is due to the author, Coulhon, Duong & Hofmann [8, Theorem 2.1] using ideas of Martell in [51] who developed a variant of the sharp function theory of Fefferman-Stein [37] in the spirit of [33], again assuming there are reasonable pointwise estimates for kernels but no regularity in the classical sense. The proof here simplifies the exposition in [8]. Shen independently proved a boundedness result similar in spirit by analogous methods [59, Theorem 3.1] which he attributes to Caffarelli and Peral [22]. In fact, it is easy to recover Shen's theorem as a consequence of this one.

for all $f \in L^2$, all ball B and all $y \in B$ where $r(B)$ is the radius of B. If $2 < p < p_0$ and $Tf \in L^p$ when $f \in L^p$ then T is strong type (p,p). More precisely, for all $f \in L^p \cap L^2$,
$$\|Tf\|_p \leq c\|f\|_p$$
where c depends only on n, p and p_0 and C.

REMARKS. (1) The operators A_r play the role of approximate identities (as $r \to 0$) eventhough $A_r(1) = 1$ is not assumed. The boundedness of A_r on L^2 is a consequence of linearity but the L^2 bounds are not explicitely needed. In applications, the L^2 bounds are uniform in r and used to check the hypotheses. The improvement in the exponents from p_0 to 2 in (1.2) and from 2 to p_0 in (1.4) expresses the regularizing effect of A_r. When $p_0 = \infty$, the left hand side of (1.4) is understood as the essential supremum on B.
(2) Possible weakening of Theorem 1.1 is as follows: the exponent 2 in (1.1) can be changed to 1 and the exponent 2 in (1.2) can be changed to $p > p_0$.
(3) As we shall see, Theorem 1.2 has little to do with operators but rather with decomposition of functions in the spirit of Fefferman-Stein's argument for the sharp function and, in fact, it is an extension of it. This is why the regularised version TA_r of T is controlled by the maximal function of $|Tf|^2$, which may be surprising at first sight.
(4) Define, for $f \in L^2$,
$$\mathcal{M}^\#_{T,A} f(x) = \sup_{B \ni x} \left(\frac{1}{|B|} \int_B |T(I - A_{r(B)})f|^2\right)^{1/2},$$
where the supremum is taken over all balls B in $\mathbb{R}^n$ containing x, and $r(B)$ is the radius of B. The assumption is that $\mathcal{M}^\#_{T,A} f$ is controlled pointwise by $(M(|f|^2))^{1/2}$. In fact, rather than the exact form of the control what matters is that $\mathcal{M}^\#_{T,A}$ is strong type (p,p) for the desired values of p.
(5) The family (A_r) indexed by positive r could be replaced by a family (A_B) indexed by balls. Then $A_{r(B)}$ is replaced by A_B in the statements. An example of such a family is given by mean values operators $A_B f = \frac{1}{|B|} \int_B f$.
(6) Note that in Theorem 1.2, T acts on L^2 but its boundedness is not needed in the proof. However, it is used in applications to check (1.4) and (1.3). Note also that T already acts on L^p and the purpose of the statement is to bound its norm. In concrete situations, this theorem is applied to suitable approximations of T, the uniformity of the bounds allowing a limiting argument to deduce L^p boundedness. So an argument to conclude for generic L^p functions is not needed here.
(7) Both theorems are valid in the vector-valued context, that is when f is valued in a Banach space B_1 and Tf is valued in another Banach space B_2. We leave to the reader the care of checking details. We apply this for square function estimates in Chapter 6.

PROOF OF THEOREM 1.1. It begins with the classical Calderón-Zygmund decomposition which we recall. [8]

---

[8] Many good references for this tool. One is [**62**].

LEMMA 1.3. *Let $n \geq 1$, $1 \leq p \leq \infty$ and $\|f\|_p < \infty$. Let $\alpha > 0$. Then, one can find a collection of cubes (Q_i), functions g and b_i such that*

(1.5) $$f = g + \sum_i b_i$$

and the following properties hold:

(1.6) $$\|g\|_\infty \leq C\alpha,$$

(1.7) $$\operatorname{supp} b_i \subset Q_i \text{ and } \int_{Q_i} |b_i|^p \leq C\alpha^p |Q_i|,$$

(1.8) $$\sum_i |Q_i| \leq C\alpha^{-p} \int_{\mathbb{R}^n} |f|^p,$$

(1.9) $$\sum_i \mathbf{1}_{Q_i} \leq N,$$

where C and N depends only on dimension and p.

Let $f \in L^{p_0} \cap L^2$. We have to prove that for any $\alpha > 0$,
$$\left|\left\{x \in \mathbb{R}^n; |Tf(x)| > \alpha\right\}\right| \leq \frac{C}{\alpha^{p_0}} \int_{\mathbb{R}^n} |f|^{p_0}.$$

Write $f = g + \sum_i b_i$ by the Calderón-Zygmund decomposition at height $\alpha > 0$. The construction of this decomposition implies that $g \in L^2$ with $\int |g|^2 \leq C\alpha^{2-p_0} \int |f|^{p_0}$. Then $Tg \in L^2$ with $\|Tg\|_2 \leq C\|g\|_2$ by the assumption in Theorem 1.1. This with (1.6) yield
$$\left|\left\{x \in \mathbb{R}^n; |Tg(x)| > \frac{\alpha}{3}\right\}\right| \leq \frac{C}{\alpha^2} \int_{\mathbb{R}^n} |g|^2 \leq \frac{C}{\alpha^{p_0}} \int_{\mathbb{R}^n} |f|^{p_0}.$$

To handle the remaining term, introduce for $r \geq 0$ the operator $B_r = I - A_r$ and let r_i be the radius of Q_i. Since
$$\left|T\left(\sum_i b_i\right)\right| \leq \sum_i |TB_{r_i} b_i| + \left|T\left(\sum_i A_{r_i} b_i\right)\right|$$

it is enough to estimate $A = |\{x \in \mathbb{R}^n; \sum |TB_{r_i} b_i(x)| > \alpha/3\}|$ and $B = |\{x \in \mathbb{R}^n; |T(\sum_i A_{r_i} b_i)(x)| > \alpha/3\}|$. Let us bound the first term. First,
$$A \leq |\cup_i 4Q_i| + \left|\left\{x \in \mathbb{R}^n \setminus \cup_i 4Q_i; \sum_i |TB_{r_i} b_i(x)| > \frac{\alpha}{3}\right\}\right|,$$

and by (1.8), $|\cup_i 4Q_i| \leq \frac{C}{\alpha^{p_0}} \int |f|^{p_0}$. To handle the other term, we observe that by Tchebytchev inequality,
$$\left|\left\{x \in \mathbb{R}^n \setminus \cup_i 4Q_i; \sum_i |TB_{r_i} b_i(x)| > \frac{\alpha}{3}\right\}\right| \leq \frac{C}{\alpha^2} \int \left|\sum_i h_i\right|^2$$

with $h_i = \mathbf{1}_{(4Q_i)^c} |TB_{r_i} b_i|$. To estimate the L^2 norm, we dualize against $u \in L^2$ with $\|u\|_2 = 1$. Write
$$\int |u| \sum_i h_i = \sum_i \sum_{j=2}^\infty A_{ij}$$

where
$$A_{ij} = \int_{C_j(Q_i)} |TB_{r_i}b_i||u|.$$

By (1.1) and (1.7),
$$\|TB_{r_i}b_i\|_{L^2(C_j(Q_i))} \le |2^{j+1}Q_i|^{1/2}g(j)\left(\frac{1}{|Q_i|}\int_{Q_i}|b_i|^{p_0}\right)^{1/p_0}$$
$$\le C|2^{j+1}Q_i|^{1/2}g(j)\alpha$$

for some appropriate constant C. Now remark that for any $y \in Q_i$ and any $j \ge 2$,
$$\left(\int_{C_j(Q_i)}|u|^2\right)^{1/2} \le \left(\int_{2^{j+1}Q_i}|u|^2\right)^{1/2} \le |2^{j+1}Q_i|^{1/2}(M(|u|^2)(y))^{1/2}.$$

Applying Hölder inequality, one obtains
$$A_{ij} \le C\alpha 2^{nj}g(j)|Q_i|(M(|u|^2)(y))^{1/2}.$$

Averaging over Q_i yields
$$A_{ij} \le C\alpha 2^{nj}g(j)\int_{Q_i}(M(|u|^2)(y))^{1/2}\,dy.$$

Summing over $j \ge 2$ and i, we have
$$\int |u|\sum_i h_i \le C\alpha \int \sum_i \mathbf{1}_{Q_i}(y)(M(|u|^2)(y))^{1/2}\,dy$$
$$\le CN\alpha \int_{\cup_i Q_i}(M(|u|^2)(y))^{1/2}\,dy$$
$$\le C'N\alpha|\cup_i Q_i|^{1/2}\||u|^2\|_1^{1/2}.$$

In the next to last inequality, we used (1.9), and in the last inequality, we used Kolmogorov's lemma and the weak type $(1,1)$ of the maximal function.[9] Hence from (1.8)
$$A \le \frac{C''N}{\alpha^{p_0}}\int_{\mathbb{R}^n}|f|^{p_0}.$$

It remains to estimate the term B. To this end, we use that T is bounded on L^2 to obtain
$$B \le \frac{C}{\alpha^2}\int\left|T\left(\sum_i A_{r_i}b_i\right)\right|^2 \le \frac{C}{\alpha^2}\int\left|\sum_i A_{r_i}b_i\right|^2.$$

To estimate the L^2 norm, we dualize against $u \in L^2$ with $\|u\|_2 = 1$ and write
$$\int |u|\sum_i |A_{r_i}b_i| = \sum_i\sum_{j=1}^\infty B_{ij}$$

where
$$B_{ij} = \int_{C_j(Q_i)}|A_{r_i}b_i||u|.$$

---

[9] This idea is borrowed from Hofmann & Martell [**42**].

Using (1.2) and (1.7),

$$\|A_{r_i} b_i\|_{L^2(F_{ij})} \leq |2^{j+1} Q_i|^{1/2} g(j) \left(\frac{1}{|Q_i|} \int_{Q_i} |b_i|^{p_0}\right)^{1/p_0}$$

$$\leq C |2^{j+1} Q_i|^{1/2} g(j) \alpha$$

for $j \geq 1$. From here, we may argue as before and conclude that B is bounded by $\frac{C}{\alpha^{p_0}} \int |f|^{p_0}$ as desired. $\square$

REMARK. As the reader can see, there is absolutely no use of mean value properties of the b_i's.

Before turning to the proof of Theorem 1.2, we state intermediate results. We begin with a useful localisation lemma.

LEMMA 1.4. *There is K_0 depending only on dimension such that the following holds. For every $f \in L^1_{loc}$ and every cube Q and every $\lambda > 0$ for which there exists $\bar{x} \in 4Q$ for which $Mf(\bar{x}) \leq \lambda$, then for every $K \geq K_0$,*

$$\{\chi_Q M f > K\lambda\} \subset \{M(f\chi_{8Q}) > \frac{K}{K_0}\lambda\}$$

PROOF. We use that M is comparable to the centered maximal function M_c: there is K_0 depending only on the doubling constant such that $M \leq K_0 M_c$.

Let $x \in Q$ with $Mf(x) > K\lambda$. Then $M_c f(x) > \frac{K}{K_0}\lambda$. Hence, there is a cube centered at x with radius r such that

$$\frac{1}{|Q(x,r)|} \int_{Q(x,r)} |f| > \frac{K}{K_0}\lambda.$$

If $\frac{K}{K_0} > 1$, $\bar{x} \notin Q(x,r)$ since $Mf(\bar{x}) \leq \lambda$. The conditions $x \in Q$, $\bar{x} \in 4Q$ and $\bar{x} \notin Q(x,r)$ imply $Q(x,r) \subset 8Q$. Hence,

$$\frac{K}{K_0}\lambda < \frac{1}{|Q(x,r)|} \int_{Q(x,r)} |f\chi_{8Q}| \leq M(f\chi_{8Q})(x).$$

$\square$

We continue with a two parameters family of good lambda inequalities.

PROPOSITION 1.5. *Fix $1 < q \leq \infty$ and $a > 1$. Then, there exist $C = C(q,n,a)$ and $K'_0 = K'_0(n,a)$ with the following property: If F, G are nonnegative measurable functions such that for every cube Q there exist non negative functions G_Q, H_Q with*

$$F \leq G_Q + H_Q \quad \text{a.e. on } Q,$$

$$\left(\frac{1}{|Q|} \int_Q H_Q^q\right)^{1/q} \leq a M F(x) \quad \text{for all } x \in Q$$

$$\frac{1}{|Q|} \int_Q G_Q \leq G(x) \quad \text{for all } x \in Q.$$

Then for all $\lambda > 0$, for all $K > K'_0$ and $\gamma < 1$,

$$|\{MF > K\lambda, G \leq \gamma\lambda\}| \leq C\left(\frac{1}{K^q} + \frac{\gamma}{K}\right) |\{MF > \lambda\}|.$$

If $q = \infty$, we understand the average in L^q as an essential supremum. In this case, $\frac{1}{K^q} = 0$.

PROOF. Let $E_\lambda = \{MF > \lambda\}$. We assume this is a proper subset in $\mathbb{R}^n$ otherwise there is nothing to prove. Since E_λ is open, the Whitney decomposition yields a family of non overlapping dyadic cubes Q_i such that $E_\lambda = \cup_i Q_i$ and $4Q_i$ contains at least one point $\overline{x_i}$ outside E_λ, that is

$$MF(\overline{x_i}) \leq \lambda.$$

Let $B_\lambda = \{MF > K\lambda, G \leq \gamma\lambda\}$. If $K \geq 1$ then $B_\lambda \subset E_\lambda$, hence

$$|B_\lambda| \leq \sum_i |B_\lambda \cap Q_i|.$$

Fix i. If $B_\lambda \cap Q_i = \emptyset$, we have nothing to do. If not, there is a point $\overline{y_i} \in Q_i$ such that

$$G(\overline{y_i}) \leq \gamma\lambda.$$

By the localisation lemma applied to F on Q_i, if $K \geq K_0$, then

$$|B_\lambda \cap Q_i| \leq |\{MF > K\lambda\} \cap Q_i| \leq |\{M(F\chi_{8Q_i}) > \frac{K}{K_0}\lambda\}|.$$

Now use $F \leq G_i + H_i$ with $G_i = G_{8Q_i}$ and $H_i = H_{8Q_i}$ to deduce

$$|\{M(F\chi_{8Q_i}) > \frac{K}{K_0}\lambda\}| \leq |\{M(G_i\chi_{8Q_i}) > \frac{K}{2K_0}\lambda\}| + |\{M(H_i\chi_{8Q_i}) > \frac{K}{2K_0}\lambda\}|.$$

Now by using the weak type $(1,1)$ and (q,q) of the maximal operator with respective constant c_1 and c_q, we have

$$|\{M(G_i\chi_{8Q_i}) > \frac{K}{2K_0}\lambda\}| \leq \frac{2K_0 c_1}{K\lambda} \int_{8Q_i} G_i$$
$$\leq \frac{2K_0 c_1}{K\lambda} |8Q_i| G(\overline{y_i})$$
$$\leq \frac{2K_0 c_1}{K} |8Q_i| \gamma,$$

and, if $q < \infty$

$$|\{M(H_i\chi_{8Q_i}) > \frac{K}{2K_0}\lambda\}| \leq \left(\frac{2K_0 c_q}{K\lambda}\right)^q \int_{8Q_i} H_i^q$$
$$\leq \left(\frac{2K_0 c_q}{K\lambda}\right)^q |8Q_i|(aMF(\overline{x_i}))^q$$
$$\leq \left(\frac{2K_0 c_q}{K}\right)^q |8Q_i| a^q.$$

Hence, summing over i yields

$$|B_\lambda| \leq \sum_i C\left(\frac{a^q}{K^q} + \frac{\gamma}{K}\right)|8Q_i| \leq C8^n \left(\frac{a^q}{K^q} + \frac{\gamma}{K}\right)|E_\lambda|.$$

If $q = \infty$, then

$$\|M(H_i\chi_{8Q_i})\|_\infty \leq \|H_i\chi_{8Q_i}\|_\infty \leq aMF(\overline{x_i}) \leq a\lambda,$$

so that, choosing $K \geq 2K_0 a$ leads us to $\{M(H_i\chi_{8Q_i}) > \frac{K}{2K_0}\lambda\} = \emptyset$. □

Let us now assume that F, G are so that the conclusion of the proposition holds. Let $0 < p < q$. Then, we have

$$\|MF\|_p^p \leq CK^p \left(\frac{a^q}{K^q} + \frac{\gamma}{K} \right) \|MF\|_p^p + \frac{K^p}{\gamma^p} \|G\|_p^p.$$

Hence, if furthermore $\|MF\|_p < \infty$, since $p < q$, one can choose K large enough and γ small enough so that

$$CK^p \left(\frac{a^q}{K^q} + \frac{\gamma}{K} \right) \leq 1 - 2^{-p}.$$

This choice depends therefore on p, q, n, a and we have

$$\|MF\|_p^p \leq \frac{(2K)^p}{\gamma^p} \|G\|_p^p.$$

PROOF OF THEOREM 1.2. Let $f \in L^2 \cap L^p$. We let $q = \frac{p_0}{2}$ and set $F = |Tf|^2 \in L^{p/2}$. By sublinearity of T, we have for any cube Q that $F \leq G_Q + H_Q$ with $G_Q = 2|T(I - A_{r(Q)})f|^2$ and $H_Q = 2|TA_{r(Q)}f|^2$. Hence the hypotheses of the proposition apply with $a = 2C^2$ and $G = 2C^2 M(f^2)$. Let $2 < p < p_0$. Since we know that $MF \in L^{p/2}$ from the hypothesis, we obtain

$$\|Tf\|_p^2 \leq \|MF\|_{p/2} \leq C\|G\|_{p/2} \leq C'\|f\|_p^2.$$

In the last inequality, we have used the assumption on $T(I - A_r)$. □

REMARK. With $q = \infty$ and given F, we let $G_Q = |F - m_Q F|$ and $H_Q = |m_Q F|$, $a = 1$ and $G = M^\# F$, the sharp function of Fefferman-Stein. We obtain if $0 < p < \infty$ and $\|MF\|_p < \infty$ that

$$\|MF\|_p \leq C_p \|M^\# F\|_p,$$

hence this argument contains in particular Fefferman-Stein's. To recover the sharp function introduced by Martell, we take $G_Q = |F - A_{r(Q)}F|$ and $H_Q = |A_{r(Q)}F|$ and, as proved Martell, $\sup_{Q \ni x} H_Q \leq aMF(x)$ under kernel upper bounds on A_r.

CHAPTER 2

Basic L^2 theory for elliptic operators

2.1. Definition

Let $A = A(x)$ be an $n \times n$ matrix of complex, L^∞ coefficients, defined on $\mathbb{R}^n$, and satisfying the ellipticity (or "accretivity") condition

(2.1) $$\lambda|\xi|^2 \leq \operatorname{Re} A\xi \cdot \overline{\xi} \text{ and } |A\xi \cdot \overline{\zeta}| \leq \Lambda|\xi||\zeta|,$$

for $\xi, \zeta \in \mathbb{C}^n$ and for some λ, Λ such that $0 < \lambda \leq \Lambda < \infty$. We define a second order divergence form operator

(2.2) $$Lf \equiv -\operatorname{div}(A\nabla f),$$

which we first interpret in the sense of maximal accretive operators via a sesquilinear form. That is, $\mathcal{D}(L)$ is the largest subspace contained in $W^{1,2}$ for which

$$\left| \int_{\mathbb{R}^n} A\nabla f \cdot \nabla g \right| \leq C\|g\|_2$$

for all $g \in W^{1,2}$ and we set Lf by

$$\langle Lf, g \rangle = \int_{\mathbb{R}^n} A\nabla f \cdot \overline{\nabla g}$$

for $f \in \mathcal{D}(L)$ and $g \in W^{1,2}$. Thus defined, L is maximal-accretive operator on L^2 and $\mathcal{D}(L)$ is dense in $W^{1,2}$. [1]

Since $W^{1,2}$ is dense in its homogeneous version $\dot{W}^{1,2}$ (for the semi-norm $\|\nabla f\|_2$), L extends to a bounded operator invertible operator from $\dot{W}^{1,2}$ into its dual space $\dot{W}^{-1,2}$, which justifies the divergence notation in (2.2). In particular, one has

(2.3) $$\|\nabla L^{-1} \operatorname{div} f\|_2 \leq c\|f\|_2.$$

2.2. Holomorphic functional calculus on L^2

Let L be as above. There exists $\omega \in [0, \frac{\pi}{2})$ depending only on the ellipticity constants such that for all $f \in \mathcal{D}(L)$,

(2.4) $$|\arg\langle Lf, f\rangle| \leq \omega.$$

We fix the smallest such ω and the following L^2 estimate is easily proved: for all $\mu \in (\omega, \pi)$ and all complex numbers $\lambda \in \Sigma_{\pi-\mu}$,

$$\|(L+\lambda)^{-1} f\|_2 \leq \frac{c\|f\|_2}{|\lambda|},$$

where we have set $\Sigma_\mu = \{z \in \mathbb{C}^*; |\arg z| < \mu\}$. [2] Hence, L is of type ω on L^2. Several consequences follow.

---

[1] For precise definitions, details and proofs, see Kato's book [46, Chapter VI].
[2] See, for example, [17, Preliminary Chapter] for a proof.

In particular, $-L$ generates a semigroup $(e^{-tL})_{t>0}$ which has an analytic extension to a complex semigroup $(e^{-zL})_{z \in \Sigma_{\frac{\pi}{2} - \omega}}$ of contractions on L^2.

Since L is also maximal-accretive operator, it has a bounded holomorphic functional calculus on L^2. In particular, for any $\mu \in (\omega, \pi)$ and any φ holomorphic and bounded in Σ_μ, the operator $\varphi(L)$ is bounded on L^2 with the estimate

$$\|\varphi(L)f\|_2 \leq c\|\varphi\|_\infty \|f\|_2,$$

the constant c depending only on ω and μ. If φ satisfies the technical condition

(2.5) $$|\varphi(\zeta)| \leq c|\zeta|^s (1+|\zeta|)^{-2s}$$

for all $\zeta \in \Sigma_\mu$ for some positive constants c, s, then $\varphi(L)$ can be computed using the semigroup. Let $\omega < \theta < \nu < \mu < \frac{\pi}{2}$. One has

(2.6) $$\varphi(L) = \int_{\Gamma_+} e^{-zL} \eta_+(z)\, dz + \int_{\Gamma_-} e^{-zL} \eta_-(z)\, dz$$

where $\Gamma_\pm$ is the half-ray $\mathbb{R}^+ e^{\pm i(\frac{\pi}{2} - \theta)}$,

(2.7) $$\eta_\pm(z) = \frac{1}{2\pi i} \int_{\gamma_\pm} e^{\zeta z} \varphi(\zeta)\, d\zeta, \quad z \in \Gamma_\pm,$$

with $\gamma_\pm$ being the half-ray $\mathbb{R}^+ e^{\pm i\nu}$ (the orientation on paths is irrelevant in the arguments where this representation is used so we do not insist on that). For general bounded holomorphic functions, $\varphi(L)$ is defined by a limiting procedure which we do not need in this work.

Finally, one can define unbounded operators $\varphi(L)$ for φ holomorphic in Σ_μ, $\omega < \mu < \pi$, satisfying

(2.8) $$|\varphi(\zeta)| \leq c \sup(|\zeta|^s, |\zeta|^{-s'})$$

for some $c, s, s' \geq 0$. This includes fractional powers of L. We call $\mathcal{F}(\Sigma_\mu)$ the class of such holomorphic functions. [3]

2.3. L^2 off-diagonal estimates

A very important ingredient in this paper is the off-diagonal estimates of Gaffney type. They are crucial to our analysis because when dealing with complex operators, we do not have at our disposal contractivity of the semigroup on all L^p spaces, and in fact this is false in general.

DEFINITION. Let $\mathcal{T} = (T_t)_{t>0}$ be a family of operators. We say that $\mathcal{T}$ satisfies L^2 off-diagonal estimates if for some constants $C \geq 0$ and $\alpha > 0$ for all closed sets E and F, all $h \in L^2$ with support in E and all $t > 0$ we have

(2.9) $$\|T_t h\|_{L^2(F)} \leq C e^{-\frac{cd(E,F)^2}{t}} \|h\|_2.$$

Here, and subsequently, $d(E, F)$ denotes the semi-distance induced on sets by the Euclidean distance.

In case $\mathcal{T} = (T_t)_{t>0}$ is replaced by a family $\mathcal{T} = (T_z)_{z \in \Sigma_\mu}$ defined on a complex sector Σ_μ with $0 \leq \mu < \frac{\pi}{2}$, then we adopt the same definition and replacing t by $|z|$ in the right hand side of the inequality. In this case, the constants C and α may depend on the angle μ.

---

[3] For definitions, we refer to [53] and [28].

2.3. L^2 OFF-DIAGONAL ESTIMATES

PROPOSITION 2.1. *For all $\mu \in (0, \frac{\pi}{2}-\omega)$, the families $(e^{-zL})_{z \in \Sigma_\mu}$, $(zLe^{-zL})_{z \in \Sigma_\mu}$ and $(\sqrt{z}\nabla e^{-zL})_{z \in \Sigma_\mu}$ satisfy L^2 off-diagonal estimates.*

PROOF. We begin with the case of real times $t > 0$. Let φ be a Lipschitz function on $\mathbb{R}^n$ with Lipschitz norm 1 and $\rho > 0$. We may define $L_\rho = e^{\rho\varphi}Le^{-\rho\varphi}$ by the form method. This operator is of second order type with same principal term as L and lower order terms with bounded coefficients. More precisely, let Q_ρ be the associated form. Then it is bounded on $W^{1,2}$ and one can find c depending only on dimension and the ellipticity constants of L (not on φ) such that

$$\Re Q_\rho(f) \geq \frac{\lambda}{2}\|\nabla f\|_2^2 - c\rho^2\|f\|_2^2, \ f \in W^{1,2}.$$

The construction guarantees that $L_\rho + c\rho^2$ is maximal-accretive on L^2. Hence, the semigroup $(e^{-tL_\rho})_{t>0}$ exists on L^2 and its analyticity gives us:

$$\|e^{-tL_\rho}f\|_2 + \|t\frac{d}{dt}e^{-tL_\rho}f\|_2 + \|\sqrt{t}\nabla e^{-tL_\rho}f\|_2 \leq Ce^{c\rho^2 t}\|f\|_2$$

for all $t > 0$ where C depends on the ellipticity constants of L and dimension only. Let E and F be two closed sets and $f \in L^2$, with compact support contained in E. Choose $\varphi(x) = d(x, E)$. Then,

(2.10) $$e^{-tL}f = e^{-\rho\varphi}e^{-tL_\rho}f.$$

Hence, for all $t > 0$, all $\rho > 0$

$$\|e^{-tL}f\|_{L^2(F)} \leq Ce^{-\rho d(E,F)}e^{c\rho^2 t}\|f\|_2.$$

Optimizing with respect to $\rho > 0$ yields

$$\|e^{-tL}f\|_{L^2(F)} \leq Ce^{-\frac{d(E,F)^2}{4ct}}\|f\|_2.$$

Next, differentiating (2.10), one has

$$\frac{d}{dt}e^{-tL}f = e^{-\rho\varphi}\frac{d}{dt}e^{-tL_\rho}f$$

and the same argument applies. Eventually, applying the gradient operator to (2.10) yields

$$\nabla e^{-tL}f = -\rho(e^{-\rho\varphi}e^{-tL_\rho}f)(\nabla\varphi) + e^{-\rho\varphi}\nabla e^{-tL_\rho}f.$$

Hence,

$$\|\nabla e^{-tL}f\|_{L^2(F)} \leq C\rho e^{-\frac{d(E,F)^2}{4ct}}\|f\|_2 + Ct^{-1/2}e^{-\rho d(E,F)}e^{c\rho^2 t}\|f\|_2$$

and choosing $\rho = \frac{d(E,F)}{2ct}$ yields

$$\|\nabla e^{-tL}f\|_{L^2(F)} \leq \frac{C}{\sqrt{t}}\left(1 + \frac{d(E,F)}{\sqrt{t}}\right)e^{-\frac{d(E,F)^2}{4ct}}\|f\|_2.$$

A density argument (since f was supposed with compact support in E) concludes the proof for real times. To go to complex times, we notice that this applies to $e^{i\alpha}L$, which is an operator in the same class as L (with coefficients $e^{i\alpha}A(x)$) as long as $|\alpha| < \frac{\pi}{2} - \omega$. Hence, the above estimates apply and the constants remain uniform when α is contained in a compact subset of $(-\frac{\pi}{2}+\omega, \frac{\pi}{2}-\omega)$. Finally, the desired estimates follow easily from the observation that $e^{-zL} = e^{-t(e^{i\alpha}L)}$ when $z = te^{i\alpha}$. □

2.4. Square root

As L is a maximal-accretive operator, it has a square root, which we denote by $L^{1/2}$, defined as the unique maximal-accretive operator such that
$$L^{1/2}L^{1/2} = L$$
as unbounded operators.[4] Many formulas can be used to compute $L^{1/2}$. The one we are going to use is
$$L^{1/2}f = \pi^{-1/2}\int_0^\infty e^{-tL}Lf \, \frac{dt}{\sqrt{t}}.$$
This equality is valid as a Bochner integral when $f \in \mathcal{D}(L)$ and as the limit in L^2 of the truncated Bochner integrals $\int_\varepsilon^R \ldots$ as $\varepsilon \downarrow 0$ and $R \uparrow \infty$ when $f \in \mathcal{D}(L^{1/2})$. Also, this construction implies that $\mathcal{D}(L)$ is dense in $\mathcal{D}(L^{1/2})$.

The determination of the domain of the square root of L has become known as the Kato square root problem and it is now a theorem in all dimensions, as recalled in the Introduction, that $\mathcal{D}(L^{1/2}) = W^{1,2}$ with
$$\|L^{1/2}f\|_2 \sim \|\nabla f\|_2 \tag{2.11}$$
for all $f \in \mathcal{D}(L)$, hence by density in $\dot{W}^{1,2}$. In particular, $L^{1/2}$ extends to an isomorphism from $\dot{W}^{1,2}$ to L^2 and the formula
$$g = L^{1/2}L^{-1/2}g$$
extends to all L^2 functions g.

This implies the following representation formula

LEMMA 2.2. *If $f, h \in \dot{W}^{1,2}$ then*
$$\langle (L^*)^{1/2}f, L^{1/2}h\rangle = \int_{\mathbb{R}^n} \nabla f \cdot \overline{A\nabla h}.$$

PROOF. Since $f, h \in \dot{W}^{1,2}$, $(L^*)^{1/2}f, L^{1/2}h \in L^2$. Hence both sides of the equality are well-defined. It suffices to obtain the equality if, in addition, $h \in \mathcal{D}(L)$, as $\mathcal{D}(L)$ is dense in $\dot{W}^{1,2}$.

In this case, $L^{1/2}h$ belongs to the domain of $L^{1/2}$, so $\langle (L^*)^{1/2}f, L^{1/2}h\rangle = \langle f, Lh\rangle$ and the latter is equal to $\int_{\mathbb{R}^n} \nabla f \cdot \overline{A\nabla h}$ by construction of L. □

2.5. The conservation property

For real operators, the semigroup is contracting on L^∞ and the conservation property $e^{-tL}1 = 1$ is a classical consequence of the probabilistic interpretation of the semigroup or of the maximum principle for parabolic equations. But for complex operators, the semigroup may not act from L^∞ into L^∞ (see the chapter on L^p theory for the semigroup). Yet, thanks to L^2 off-diagonal estimates, the action of the semigroup on L^∞ can be defined in the L^2_{loc} sense and the conservation property still holds in this sense:[5]
$$e^{-tL}1 = 1, \quad t > 0. \tag{2.12}$$

---

[4] For an explicit construction, we refer the reader to Kato's book [**46**, p. 281] or to Meyer and Coifman's book [**56**, Chapter XIV].

[5] This is proved under $L^1 - L^\infty$ off diagonal estimates of the semigroup in [**17**] (Chapter I, Proposition 25) and is a consequence of Corollary 4.6 in [**7**] under the weaker L^∞-boundedness. It is mentioned in all generality but without proof in [**10**], p. 638.

The first step is to show that e^{-tL} maps L^2 functions with compact supports in L^1. Fix $t > 0$ and $\phi \in L^2$ supported in a cube Q. Cover $\mathbb{R}^n$ with a family of nonoverlapping cubes (Q_k) of constant size with $Q_0 \supset 2Q$. Using the L^2 off-diagonal estimates,

$$\int_{Q_k} |e^{-tL}\phi| \leq |Q_k|^{1/2} e^{-\frac{cd(Q_k,Q)^2}{t}} \|\phi\|_2 \tag{2.13}$$

so that summing in k gives us the result.

Applying the first step to L^* means that one can define $e^{-tL}1$ in L^2_{loc} by

$$\int_{\mathbb{R}^n} e^{-tL}1 \,\overline{\phi} = \int_{\mathbb{R}^n} \overline{e^{-tL^*}\phi}$$

for all ϕ in L^2 with compact support.

Next, let $\mathcal{X}$ be a smooth function with $\mathcal{X}(x) = 1$ if $|x| \leq 1$ and $\mathcal{X}(x) = 0$ if $|x| \geq 2$. Let $\mathcal{X}_R(x) = \mathcal{X}(x/R)$ for $R > 0$. If ϕ is an L^2 compactly supported function, then for $R > 0$ and $t > 0$

$$\int_{\mathbb{R}^n} e^{-tL}1\,\overline{\phi} = \int_{\mathbb{R}^n} \mathcal{X}_R \overline{e^{-tL^*}\phi} + \int_{\mathbb{R}^n} (1-\mathcal{X}_R)\overline{e^{-tL^*}\phi}. \tag{2.14}$$

We use this representation twice, first to show that the left hand side does not depend on $t > 0$ and, second, to find $\int_{\mathbb{R}^n} \overline{\phi}$ as its value. This, indeed, shows that $e^{-tL}1 = 1$ in the sense of L^2_{loc}.

Let us begin with differentiating (2.14) with respect to t. Indeed, the first step applies also to $\frac{d}{dt}e^{-tL^*}$ by the L^2 off-diagonal estimates and this allows us to use the Lebesgue differentiation theorem to see that

$$\frac{d}{dt}\int_{\mathbb{R}^n} e^{-tL}1\,\overline{\phi} = \int_{\mathbb{R}^n} \mathcal{X}_R \overline{\frac{d}{dt}e^{-tL^*}\phi} + \int_{\mathbb{R}^n} (1-\mathcal{X}_R)\overline{\frac{d}{dt}e^{-tL^*}\phi}.$$

Fix $t > 0$ and let $R \to \infty$. By Lebesgue dominated convergence, the latter integral tends to 0. Now, since $\nabla \mathcal{X}_R \in L^2$ and $\frac{d}{dt}e^{-tL^*}\phi \in \mathcal{D}(L^*)$, we have that

$$\int_{\mathbb{R}^n} \mathcal{X}_R \overline{\frac{d}{dt}e^{-tL^*}\phi} = \int_{\mathbb{R}^n} A\nabla \mathcal{X}_R \cdot \overline{\nabla e^{-tL^*}\phi}.$$

Again, using the L^2 off-diagonal estimates for ∇e^{-tL^*} and arguing as for (2.13), this integral is bounded by $CR^{n/2-1}e^{-cR^2/t}\|\phi\|_2$ for R large enough so that it tends to 0. This shows that the left hand side of (2.14) is independent of $t > 0$. In the right hand side, choose and fix R large enough so that the supports of ϕ and $(1-\mathcal{X}_R)$ are far apart. It follows from (2.13) that $\int_{\mathbb{R}^n}(1-\mathcal{X}_R)\overline{e^{-tL^*}\phi}$ tends to 0 with t by dominated convergence. Eventually, since e^{-tL^*} is a strongly continuous in L^2 at $t = 0$, we obtain that $\int_{\mathbb{R}^n} \mathcal{X}_R \overline{e^{-tL^*}\phi}$ tends to $\int_{\mathbb{R}^n} \mathcal{X}_R \overline{\phi} = \int_{\mathbb{R}^n} \overline{\phi}$ as t tends to 0. This proves (2.12).

CHAPTER 3

L^p theory for the semigroup

This chapter is devoted to establishing the basic properties concerning uniform boundedness of the semigroup $(e^{-tL})_{t>0}$ and of the family $(\sqrt{t}\,\nabla e^{-tL})_{t>0}$ on L^p spaces.

3.1. Hypercontractivity and uniform boundedness

The point of this section is to present a general statement allowing to pass from hypercontractivity properties for the semigroup to uniform boundedness properties. The bridge between both are off-diagonal estimates. [1]

We introduce a few definitions.

DEFINITION. Let $\mathcal{T} = (T_t)_{t>0}$ be a family of uniformly bounded operators on L^2. We say that $\mathcal{T}$ is $L^p - L^q$ bounded for some $p, q \in [1, \infty]$ with $p \leq q$ if for some constant C, for all $t > 0$ and all $h \in L^p \cap L^2$

(3.1) $$\|T_t h\|_q \leq C t^{-\gamma_{pq}/2} \|h\|_p.$$

We say that $\mathcal{T}$ satisfies $L^p - L^q$ off-diagonal estimates for some $p, q \in [1, \infty]$ with $p \leq q$ if if for some constants $C, c > 0$, for all closed sets E and F, all $h \in L^p \cap L^2$ with support in E and all $t > 0$ we have

(3.2) $$\|T_t h\|_{L^q(F)} \leq C t^{-\gamma_{pq}/2} e^{-\frac{cd(E,F)^2}{t}} \|h\|_p.$$

Recall that the numbers γ_{pq} are defined in Notation. Such estimates depend on dimension and on the "parabolic" character of the family through this number. If $p = q$ we speak of L^p boundedness and L^p off-diagonal estimates. Remark that these notions dualize. A family $(T_t)_{t>0}$ is $L^p - L^q$ bounded (resp. satisfies $L^p - L^q$ off-diagonal estimates if and only if the dual family $(T_t^*)_{t>0}$ is $L^{q'} - L^{p'}$ bounded (resp. satisfies $L^{q'} - L^{p'}$ off-diagonal estimates).

Let us also state a useful result whose easy proof is skipped.

PROPOSITION 3.1. *If $(T_t)_{t>0}$ satisfies $L^p - L^q$ boundedness (resp. off-diagonal estimates) and $(S_t)_{t>0}$ satisfies $L^q - L^r$ boundedness (resp. off-diagonal estimates) then $(S_t T_t)_{t>0}$ satisfies $L^p - L^r$ boundedness (resp. off-diagonal estimates).*

For a semigroup, the terminology **hypercontractivity** is often used for $L^p - L^q$ boundedness for some $p < q$. The relation between hypercontractivity, boundedness and off-diagonal estimates is the following result.

PROPOSITION 3.2. *Let $p \in [1, 2)$ and $n \geq 1$. Let $\mathcal{S} = (e^{-tL})_{t>0}$.*
  (1) If $\mathcal{S}$ is L^p bounded then it is $L^p - L^2$ bounded.

---

[1] This tool was introduced for that purpose by Davies [29]. See [50] for further historical comments. This notion turns out to be equivalent to some $L^p - L^q$ boundedness for perturbed semigroups by exponential weights (See [50]). We do not need to go into this here.

15

(2) If $\mathcal{S}$ is $L^p - L^2$ bounded, then for all $q \in (p, 2)$ it satisfies $L^q - L^2$ off-diagonal estimates.

(3) If $\mathcal{S}$ satisfies $L^p - L^2$ off-diagonal estimates then it is L^p bounded.

REMARK. The result applies when $2 < p \leq \infty$ by duality: replace $L^p - L^2$ by $L^2 - L^p$ everywhere. We have privileged the central role of L^2 for reasons of simplicity and usefulness. L^2 could be replaced by L^q for q larger than 2 if necessary. It occurs in Section 4.5.

PROOF. The proof of item 1 is obtained from Nash type inequalities.[2] We start from the Gagliardo-Nirenberg inequality

$$\|f\|_2^2 \leq C \|\nabla f\|_2^{2\alpha} \|f\|_p^{2\beta}$$

with

$$\alpha + \beta = 1 \quad \text{and} \quad (1 + \gamma_p)\alpha = \gamma_p.$$

This yields the Nash inequality

$$\|e^{-tL} f\|_2^2 \leq C \|\nabla e^{-tL} f\|_2^{2\alpha} \|e^{-tL} f\|_p^{2\beta}$$

for all $t > 0$ and $f \in L^2 \cap L^p$. By ellipticity, one has

$$\|\nabla e^{-tL} f\|_2^2 \leq \lambda^{-1} \Re \int_{\mathbb{R}^n} A \nabla e^{-tL} f \cdot \overline{\nabla e^{-tL} f} = -(2\lambda)^{-1} \frac{d}{dt} \|e^{-tL} f\|_2^2.$$

Assume $f \in L^2 \cap L^p$ with $\|f\|_p = 1$. Using L^p boundedness of the semigroup in the Nash inequality, one obtains the differential inequality

$$\varphi(t)^{1/\alpha} \leq -C \varphi'(t),$$

where $\varphi(t) = \|e^{-tL} f\|_2^2$. Integrating between t and $2t$ and using that φ is nonincreasing, one finds easily that

$$\varphi(t) \leq C t^{-\frac{\alpha}{\alpha-1}} = C t^{-\gamma_p},$$

which is the desired estimate.

The proof of item 2 consists in interpolating by the Riesz-Thorin theorem the $L^p - L^2$ boundedness assumption with the L^2 off-diagonal estimates, once we fix the sets E and F in the definition of the off-diagonal estimates.

The proof of item 3 can be seen by invoking the following simple lemma which has nothing to do with semigroups.

LEMMA 3.3.[3] If $1 \leq p \leq q \leq \infty$ and T is a linear operator which satisfies $L^p - L^q$ off-diagonal estimates in the form $\|Tf\|_{L^q(F)} \leq g(d(E,F)) \|f\|_{L^p(E)}$ whenever E, F are closed cubes and f is supported in E and g is some function. Then T is bounded on L^p with norm bounded by $s^{\gamma_{pq}} \sum_{k \in \mathbb{Z}^n} g(\sup(|k|-1, 0)s)$ for any $s > 0$ provided this sum is finite.

---

[2] The proof follows that of [65, Theorem II.3.2] done for $p = 1$.
[3] This result is implicit in the proof of [29, Theorem 25].

PROOF. We may assume $p \leq q < \infty$. Let $(Q_k)_{k\in\mathbb{Z}^n}$ be a partition of $\mathbb{R}^n$ by cubes having sidelength s and $\mathcal{X}_\ell$ be the indicator function of Q_ℓ. Then

$$\|Tf\|_q^q = \sum_k \left\|\sum_\ell T(f\mathcal{X}_\ell)\right\|_{L^q(Q_k)}^q$$

$$\leq \sum_k \left(\sum_\ell g(\sup(|k-\ell|-1,0)s)\|f\|_{L^p(Q_\ell)}\right)^q$$

$$\leq \sum_k \left(\sum_\ell g(\sup(|k-\ell|-1,0)s)s^{\gamma_{pq}}\|f\|_{L^q(Q_\ell)}\right)^q$$

$$\leq \left(\sum_k g(\sup(|k-\ell|-1,0)s)s^{\gamma_{pq}}\right)^q \sum_\ell \|f\|_{L^q(Q_\ell)}^q$$

$$= \left(\sum_k g(\sup(|k|-1,0)s)s^{\gamma_{pq}}\right)^q \|f\|_q^q$$

where we used that the discrete convolution with an ℓ^1 sequence is bounded on ℓ^q. □

We come back to the proof of the proposition. It suffices to apply this lemma to $T = e^{-tL}$ from L^p to L^2 with $g(u) = ct^{-\gamma_p/2}e^{-cu^2/t}$ and choose $s = t^{1/2}$. This yields L^p boundedness of $\mathcal{S}$. □

REMARK. For the **same** p, $L^p - L^2$ off-diagonal estimates $\Longrightarrow L^p$ boundedness $\Longrightarrow L^p - L^2$ boundedness for the semigroup $\mathcal{S}$. We do not know the status of the converses for this class of semigroups. It would be of great interest when $p = 1$.

3.2. $W^{1,p}$ elliptic estimates and hypercontractivity

In this section, we show how to obtain hypercontractivity properties using $W^{1,p}$ elliptic estimates. We proceed independently of dimension although specific arguments for dimensions 1 and 2 yield much better results.

LEMMA 3.4. [4] *There is an $r \in [1,2)$ depending on dimension and the ellipticity constants only, such that $I + L$ extends to a bounded and invertible operator from $W^{1,p}$ onto $W^{-1,p}$ for $|\frac{1}{2} - \frac{1}{p}| < |\frac{1}{2} - \frac{1}{r}|$.* [5]

PROOF. That $I + L$ is bounded from $W^{1,p}$ into $W^{-1,p}$ for all p with $1 \leq p \leq \infty$ is obvious.

Let A denote the operator of pointwise multiplication with $A(x)$ and $\|A\|$ its norm acting on L^p spaces of $\mathbb{C}^n$-valued functions (endowing matrices with the norm inherited from the Hermitian structure on $\mathbb{C}^n$). This is the same number for all $1 \leq p \leq \infty$. By ellipticity, there exists a large constant c such that $\|A - cI\| < c$. Thus, $A = c(I + M)$ where $\|M\| < 1$. This means that

$$L = (I - c\Delta - c\operatorname{div} M\nabla) = J(I - J^{-1}\operatorname{div} M\nabla J^{-1})J,$$

---

[4] In this generality, this is [**15**, Proposition 3.1]. See also [**19**] and [**44**]. Note that when $n = 1$, invertibility holds for $1 \leq p \leq \infty$ [**15**, Theorem 2.2]. If $n \geq 2$, r may be arbitrary close to 2.

[5] There is also the corresponding homogeneous statement for L from $\dot{W}^{1,p}$ onto $\dot{W}^{-1,p}$. See Section 4.5.

where Δ denotes the ordinary Laplacian and $J = c^{-1/2}(I - c\Delta)^{1/2}$.

By standard multiplier theorems (or kernel estimates and Calderón-Zygmund theory) the array of operators ∇J^{-1} is bounded from L^p (of $\mathbb{C}$-valued functions) to L^p (of $\mathbb{C}^n$-valued functions) for $1 < p < \infty$ ($p = 1$ and $p = \infty$ included if $n = 1$). Moreover, a Fourier transform argument shows that $c_2 = 1$ if c_p denotes the norm on L^p.

Thus $R = J^{-1} \operatorname{div} M \nabla J^{-1}$ is bounded on L^p for $1 < p < \infty$ with norm bounded above by $\|M\| c_p^2$. Since c_p is controlled by a convex function, the operator norm of R on L^p remains less than 1 provided p is close to 2. Therefore, one can invert $I - R$ by a converging Neumann series in the space bounded operators on L^p for p close to 2. Since J is bounded and invertible from $W^{s,p}$ onto $W^{s-1,p}$ for $s = 0, 1$ and $1 < p < \infty$, this proves the invertibility of $I + L$ from $W^{1,p}$ onto $W^{-1,p}$ for p close to 2. $\square$

COROLLARY 3.5. *Let* $r_* = \frac{nr}{n+r}$ *with r as in Lemma 3.4. Let* $p \in [1,2]$ *be such that*
$$\begin{cases} p > r_*, & \text{if } r_* \geq 1 \\ p = 1 & \text{if } r_* < 1. \end{cases}$$
The semigroup $(e^{-tL})_{t>0}$ *and its dual* $(e^{-tL^*})_{t>0}$ *are* $L^p - L^2$ *bounded and the best constant C in*

(3.3) $$\|e^{-tL}f\|_2 + \|e^{-tL^*}f\|_2 \leq C t^{-\gamma_p/2} \|f\|_p, \quad f \in L^2 \cap L^p$$

depends only on dimension, ellipticity and p.

PROOF. Assume first that $t = 1$. By Lemma 3.4 and the Sobolev embedding theorem, in a finite number of steps $(1 + L)^{-k}$ extends to a bounded map from L^p into L^2. Note that k depends only on r, hence ellipticity and dimension. Let $f \in L^2 \cap L^p$. Since f is in L^2, the equality
$$e^{-L}f = e^{-L}(I + L)^k (I + L)^{-k} f$$
is justified. As $e^{-L}(I + L)^k$ extends to bounded operator on L^2 by analyticity of the semigroup on L^2, we have obtained that $\|e^{-L}f\|_2 \leq C\|f\|_p$, with a constant C that depends only on ellipticity, dimension and p.

If $t \neq 1$, then the affine change of variable in $\mathbb{R}^n$ defined by $g(x) = f(t^{1/2}x)$ gives us $e^{-tL}f(x) = (e^{-L_t}g)(t^{-1/2}x)$ with L_t the second order operator associated with the matrix of coefficients $A(t^{1/2}x)$. Since L_t has same ellipticity constants as L, the previous bound applies and yields the desired estimate in function of t by change of variables.

The same argument applies to L^*. $\square$

COROLLARY 3.6. [6] *If $n = 1$ or 2, then the semigroup* $(e^{-tL})_{t>0}$ *is* $L^p - L^2$ *bounded for $1 \leq p < 2$ and $L^2 - L^p$ bounded for $2 < p \leq \infty$. Hence, it is L^p-bounded for all $p \in (1, \infty)$. If $n \geq 3$, there exists $\varepsilon > 0$ depending only on dimension and the ellipticity constants such that the semigroup* $(e^{-tL})_{t>0}$ *is L^p-bounded for all p contained in the interval* $(p_n - \varepsilon, (p_n - \varepsilon)')$ *for* $p_n = \frac{2n}{n+2}$.

---

[6] For $n = 1$, this is in [15], Theorem 2.21, for $n = 2$ in [15], Theorem 3.5. There, L^1 and L^∞ boundedness and even $L^1 - L^\infty$ off-diagonal estimates are proved, but further arguments such as the so-called Davies' trick are needed. This method, sufficient for our needs here, is not powerful enough. For $n \geq 3$, this follows from combining the works of Davies [29] for $p_n \leq p \leq (p_n)'$ and the perturbation method in [17], Chapter I.

PROOF. In dimensions $n = 1, 2$, we have $r_* < 1$. In dimension $n \geq 3$, the value of r_* may or may not be less than 1, but it is less than $2_* = \frac{2n}{n+2}$. It suffices to combine Corollary 3.5 and Proposition 3.2 to finish the proof. □

We introduce here two critical exponents for L. Let $\mathcal{J}(L)$ denote the maximal interval of exponents p in $[1, \infty]$ for which the semigroup $(e^{-tL})_{t>0}$ is L^p bounded.[7] We write $\operatorname{Int} \mathcal{J}(L) = (p_-(L), p_+(L))$. Note that $(p_+(L))' = p_-(L^*)$ and vice versa. One has shown $p_+(L) = \infty$ and $p_-(L) = 1$ if $n = 1, 2$, and $p_+(L) > \frac{2n}{n-2}$ and $p_-(L) < \frac{2n}{n+2}$ if $n \geq 3$. In specific situations, much more can be said. For example, we have from the well-known formula for the heat kernel $p_-(-\Delta) = 1$, $p_+(-\Delta) = \infty$. From the maximum principle for real parabolic equations, one has $p_-(L) = 1$, $p_+(L) = \infty$ if L has real coefficients.

3.3. Gradient estimates

Let us consider the possible estimates for ∇e^{-tL}. Let $\mathcal{N}(L)$ denote the maximal interval (if nonempty) of exponents p in $[1, \infty]$ for which the family $(\sqrt{t}\, \nabla e^{-tL})_{t>0}$ is L^p bounded. We write $\operatorname{Int} \mathcal{N}(L) = (q_-(L), q_+(L))$. A dichotomy between the cases $p > 2$ and $p < 2$ appears immediatly.

PROPOSITION 3.7. *Let* $1 \leq p < 2$. *If* $(\sqrt{t}\, \nabla e^{-tL})_{t>0}$ *is* L^p-*bounded, then* $(e^{-tL})_{t>0}$ *is* L^q-*bounded for* $p < q < 2$. *Conversely, if* $(e^{-tL})_{t>0}$ *is* L^p-*bounded then* $(\sqrt{t}\, \nabla e^{-tL})_{t>0}$ *is* L^q-*bounded for* $p < q < 2$. *Hence* $q_-(L) = p_-(L)$.

PROOF. Let us see the direct part first. We have nothing to prove if $n \leq 2$ by Corollary 3.6 as the conclusion holds true. If $n \geq 3$, by interpolation and Sobolev embeddings we have that $(e^{-tL})_{t>0}$ is $L^q - L^{q^*}$ bounded for $p \leq q \leq 2$. Let p_k be defined by $p_0 = p$, $p_{k+1} = (p_k)^*$ and stop whenever $2_* \leq p_k < 2$. By composition and the semigroup property, we have that $(e^{-tL})_{t>0}$ is $L^p - L^{p_k}$ bounded. By Corollary 3.5, since $p_k \geq 2_*$, $(e^{-tL})_{t>0}$ is $L^{p_k} - L^2$ bounded. By composition again, $(e^{-tL})_{t>0}$ is $L^p - L^2$ bounded. The conclusion follows from Proposition 3.2.

For the converse, by Proposition 3.2, $(e^{-tL})_{t>0}$ is $L^p - L^2$ bounded. Since $(\sqrt{t}\, \nabla e^{-tL})_{t>0}$ is L^2 bounded, this self-improves by composition to $L^p - L^2$ boundedness. Lemma 3.3 applied to $T = \nabla e^{-tL}$ yields the conclusion.

The relation $q_-(L) = p_-(L)$ follows immediately. □

REMARK. (1) We see that the semigroup acting on L^p self-improves into $W^{1,p}$ if $p < 2$ up to allowing the p's to vary. This is false when $p > 2$.
(2) Although we do not need such a refinement here, it would be interesting to know whether the conclusions of L^p boundedness hold at the endpoint p. The arguments show, nevertheless, that, for the **same** p, $L^p - L^2$ boundedness for $(e^{-tL})_{t>0}$ is equivalent to $L^p - L^2$ boundedness for $(\sqrt{t}\, \nabla e^{-tL})_{t>0}$, and $L^p - L^2$ off-diagonal estimates for $(e^{-tL})_{t>0}$ implies $L^p - L^2$ off-diagonal estimates for $(\sqrt{t}\, \nabla e^{-tL})_{t>0}$. The converse of the latter is not clear.[8]

Let us record here the following consequences of the above argument for later use.

---

[7] In [50], a systematic study of this interval is made for a class of real elliptic operators. The methods heavily rely on real functions.

[8] In recent work with J.-M. Martell, we established the converse [12].

COROLLARY 3.8. *Let $p \in [1, 2)$ and $n \geq 1$. Let $\mathcal{N} = (\sqrt{t}\, \nabla e^{-tL})_{t>0}$.*
1. *If $\mathcal{N}$ is L^p bounded then it is $L^p - L^2$ bounded.*
2. *If $\mathcal{N}$ is $L^p - L^2$ bounded, then it satisfies $L^q - L^2$ off-diagonal estimates for $p < q < 2$.*
3. *If $\mathcal{N}$ satisfies $L^p - L^2$ off-diagonal estimates then it is L^p bounded.*

Next, let us consider the case $p > 2$. We have the same statement as the corollary but with a different argument.

PROPOSITION 3.9. *Let $p \in (2, \infty]$ and $n \geq 1$. Let $\mathcal{N} = (\sqrt{t}\, \nabla e^{-tL})_{t>0}$.*
1. *If $\mathcal{N}$ is L^p bounded then it is $L^2 - L^p$ bounded.*
2. *If $\mathcal{N}$ is $L^2 - L^p$ bounded, then it satisfies $L^2 - L^q$ off-diagonal estimates for $2 < q < p$.*
3. *If $\mathcal{N}$ satisfies $L^2 - L^p$ off-diagonal estimates then it is L^p bounded.*

PROOF. Let us prove the first item. Assume that $n \geq 3$ and also that $p < \infty$. If ∇e^{-tL} is bounded on L^p with bound $Ct^{-1/2}$, then the same is true for all $q \in [2, p]$. By Sobolev embeddings, for all $q \in [2, p]$ with $q < n$,
$$e^{-tL} \colon L^q \to L^{q^*}$$
with bound $Ct^{-1/2}$. Since we know that $e^{-tL} \colon L^2 \to L^q$ for any $q \in \mathbb{R}$ with $2 \leq q \leq 2^*$ from Corollary 3.5, it follows as in the proof of Proposition 3.7 that e^{-tL} maps L^2 to L^q for all $q \in \mathbb{R}$ with $2 \leq q \leq p^*$. Writing ∇e^{-tL} as $\nabla e^{-(t/2)L} e^{-(t/2)L}$ shows that this operator is bounded from L^2 into L^p with the appropriate norm growth.

Assume next $n \geq 3$ and $p = \infty$. We just need to prove that the semigroup is $L^2 - L^\infty$ bounded and the rest of the argument applies. By Morrey's embedding, if $n < q < \infty$ and $\alpha = 1 - \frac{n}{q}$, $t > 0$ and $x, y \in \mathbb{R}^n$,
$$|e^{-tL}f(x) - e^{-tL}f(y)| \lesssim \|\nabla e^{-tL}f\|_q |x-y|^\alpha \lesssim t^{-\frac{1}{2} - \frac{\gamma_q}{2}} \|f\|_2 |x-y|^\alpha.$$
In the last inequality, we used that $\mathcal{N}$ is $L^2 - L^q$ bounded as we have just proved it. Fix x and average the square of this inequality on a ball B with center x and radius $r = \sqrt{t}$ to find
$$|e^{tL}f(x)| \lesssim |B|^{-1/2} \|e^{-tL}f\|_{L^2(B)} + r^{\frac{\alpha}{2}} t^{-\frac{1}{2} - \frac{\gamma_q}{2}} \|f\|_2 \lesssim t^{-n/4} \|f\|_2$$
by using the L^2 boundedness of e^{-tL}. This proves the $L^2 - L^\infty$ boundedness of the semigroup.

If $n \leq 2$, we already know that $e^{-tL} \colon L^2 \to L^q$ for any $q \in [2, \infty]$ from Corollary 3.5 and the rest of the above argument applies.

The second item is a consequence of interpolation between the hypothesis and the L^2 off-diagonal estimates for $(\sqrt{t}\, \nabla e^{-tL})_{t>0}$.

For the third item, it is enough to apply Lemma 3.3 to $T = e^{-tL^*} \operatorname{div}$ with $s = t^{1/2}$ and duality. □

COROLLARY 3.10. *$q_+(L) > 2$.*

PROOF. Let $p > 2$ such that $|\frac{1}{2} - \frac{1}{p}| < |\frac{1}{2} - \frac{1}{r}|$ where r is given in Lemma 3.4. Let $f \in L^2$. After a finite number of steps (depending only on r) $(I+L)^{-k}$ maps L^2 into $W^{1,p}$. In particular, since $(I+L)^k e^{-L} f \in L^2$ by analyticity,
$$\nabla e^{-L} f = \nabla (I+L)^{-k} (I+L)^k e^{-L} f \in L^p.$$

This proves the boundedness of ∇e^{-L} from L^2 into L^p. The L^2-L^p boundedness of the family $(\sqrt{t}\,\nabla e^{-tL})_{t>0}$ follows by the rescaling argument of the proof of Corollary 3.5. Hence, $q_+(L) \geq p$. □

COROLLARY 3.11. *$\mathcal{N}(L)$ is not empty and contains a neighborhood of 2.*

PROOF. We have $q_-(L) = p_-(L) < 2$ and $q_+(L) > 2$. □

More is true in dimension 1.

PROPOSITION 3.12. *If $n = 1$, we have $q_+(L) = \infty$.*

PROOF. In dimension 1, the operator L takes the form $-\frac{d}{dx}\left(a\frac{d}{dx}\right)$. By Proposition 3.9, it suffices to establish that $(\sqrt{t}\,\frac{d}{dx}e^{-tL})_{t>0}$ is $L^2 - L^\infty$ bounded. [9] Let $f \in L^2$ and $t > 0$ and set $g = \sqrt{t}\,a\frac{d}{dx}e^{-tL}f$. We know that $g \in L^2$ with norm $O(1)$, while $g' = -\sqrt{t}\,Le^{-tL}f \in L^\infty$ with norm $O(t^{-1/2})$. The last fact can be seen from writing $Le^{-tL} = e^{-(t/2)L}Le^{-(t/2)L}$ and using that $(tLe^{-tL})_{t>0}$ is L^2 bounded and $(e^{-tL})_{t>0}$ is $L^2 - L^\infty$ bounded by Corollary 3.6. This implies that $g \in L^\infty$ with norm $O(t^{-1/4})$. Indeed, let I be an interval of size $t^{1/4}$ and $x, y \in I$. We have $g(x) - g(y) = \int_y^x g'(s)\,ds$. Hence, $|g(x)| \leq |g(y)| + Ct^{-1/4}$. Averaging squares over I with respect to y yields $|g(x)|^2 \leq Ct^{-1/2}\|g\|_2^2 + Ct^{-1/2} = O(t^{-1/2})$. Since I and x are arbitrary, this gives us the desired L^∞ bound on $\mathbb{R}$. □

PROPOSITION 3.13. *Assume $2 < p \leq \infty$ and $(\sqrt{t}\,\nabla e^{-tL})_{t>0}$ is L^p-bounded.*
(1) *$(e^{-tL})_{t>0}$ is L^q-bounded for $2 \leq q \leq p^*$ except when $p = n$ for which we conclude only for $2 \leq q < \infty$. In particular, one has $p_+(L) \geq (q_+(L))^*$.*
(2) *If $p < \infty$, $(t\nabla e^{-tL}\,\mathrm{div})_{t>0}$ is L^p-bounded.*

PROOF. The first part has been seen in the proof of Proposition 3.9. For the second part, the L^p boundedness of $(\sqrt{t}\,\nabla e^{-tL})_{t>0}$ implies in particular that $(e^{-tL})_{t>0}$ is L^q bounded for q in a neighborhood of p, thus, that $(e^{-tL^*})_{t>0}$ is L^q bounded for q in a neighborhood of p'. Applying Proposition 3.7 shows, in particular, that $(\sqrt{t}\,\nabla e^{-tL^*})_{t>0}$ is $L^{p'}$-bounded, hence, that $(\sqrt{t}\,e^{-tL}\,\mathrm{div})_{t>0}$ is L^p-bounded by duality. We conclude by composition. □

REMARK. If $n = 1$, combining Corollary 3.6, Proposition 3.12 and Proposition 3.13, one has that the semigroup is bounded from $W^{-1,p}$ into $W^{1,p}$ for $1 < p < \infty$ and $t > 0$. [10]

3.4. Summary

This section is nothing but a summary of results obtained so far and gathered here for the reader's convenience.

Recall that $\mathcal{J}(L)$ denotes the maximal interval of exponents p in $[1, \infty]$ for which the semigroup $(e^{-tL})_{t>0}$ is L^p bounded. We write $\mathrm{Int}\,\mathcal{J}(L) = (p_-(L), p_+(L))$. Note that $(p_+(L))' = p_-(L^*)$ and vice versa. One has $p_+(L) = \infty$ and $p_-(L) = 1$ if $n = 1, 2$ and $p_+(L) > \frac{2n}{n-2}$ and $p_-(L) < \frac{2n}{n+2}$ if $n \geq 3$.

---

[9] In fact, this family satisfies the $L^1 - L^\infty$ off-diagonal estimates, which implies all the other boundedness properties. See [18, p. 728] where it is mentioned, and [15, Theorem 2.21] for details.

[10] More is true, in fact, this holds for the endpoints $p = 1$ and $p = \infty$ and one even has $L^1 - L^\infty$ off-diagonal estimates for the family $(t\frac{d}{dx}e^{-tL}\frac{d}{dx})_{t>0}$. See [AMcT].

Next, $\mathcal{N}(L)$ denotes the maximal interval of exponents p in $[1,\infty]$ for which the family $(\sqrt{t}\,\nabla e^{-tL})_{t>0}$ is L^p bounded. We write $\operatorname{Int}\mathcal{N}(L) = (q_-(L), q_+(L))$. We have $q_-(L) = 1$ and $q_+(L) = \infty$ if $n = 1$, $q_-(L) = 1$ if $n = 2$, $q_-(L) < \frac{2n}{n+2}$ if $n \geq 3$ and $q_+(L) > 2$ if $n \geq 2$. In particular, $\operatorname{Int}\mathcal{N}(L)$ is a neighborhood of 2.

The relation between $p_\pm(L)$ and $q_\pm(L)$ are as follows:
$$q_-(L) = p_-(L),$$
$$p_+(L) \geq (q_+(L))^*.$$

3.5. Sharpness issues

Denote by $\mathcal{E}$ the class of all complex elliptic operators under study.

It is known that if $n \geq 5$, there exists $L \in \mathcal{E}$ for which $p_-(L) > 1$. [11]

It is not known if the inequality $p_-(L) < \frac{2n}{n+2}$ ($n \geq 3$), or equivalently by taking the adjoint $p_+(L) > \frac{2n}{n-2}$ is sharp: that is, if given $p < \frac{2n}{n+2}$, there exists $L \in \mathcal{E}$ for which $p_-(L) > p$.

In view of this a natural conjecture is

CONJECTURE 3.14. *If $n \geq 3$, the inequality $p_-(L) < \frac{2n}{n+2}$ is sharp for the class $\mathcal{E}$.*

It is known that the inequality $q_+(L) > 2$ is sharp in dimensions $n \geq 2$: for any $\varepsilon > 0$, there exists $L \in \mathcal{E}$ (in fact, L is real and symmetric) such that $q_+(L) \leq 2 + \varepsilon$. [12] As a consequence, there is no general upper bound of $p_+(L)$ in terms of $q_+(L)$ since $p_+(L) = \infty$ if $n = 2$.

It is not known whether the inequality $p_+(L) \geq q_+(L)^*$ is best possible ($n \geq 3$): that is, if given $\varepsilon > 0$, one can find $L \in \mathcal{E}$ with $p_+(L) < q_+(L)^* + \varepsilon$.

3.6. Analytic extension

For technical reasons, we often need to apply the above results to the analytic extension of the semigroup associated to L. We come to this now.

The definition 3.1 of $L^p - L^q$ boundedness and $L^p - L^q$ off-diagonal estimates applies to a family $\mathcal{T} = (T_z)_{z \in \Sigma_\beta}$ defined on a complex sector Σ_β with $0 \leq \beta < \frac{\pi}{2}$, replacing t by $|z|$ in the right hand side of the inequalities (3.1) and (3.2).

Proposition 3.1 extends right away to such complex families. The statement of Proposition 3.2 is also true for the analytic extension of the semigroup. However, more is true. It suffices to know boundedness of the semigroup to obtain all properties for its analytic extension with optimal angles of the sectors, that is, the sectors are p-**independent**. [13]

---

[11] This is proved in [9], based on counterexamples built in [52]. See [17] for a shorter argument due to Davies.

[12] This follows from Meyers' example. See [17].

[13] The history of p-independence of sector of holomorphy for $1 \leq p < \infty$ begins with that of p-independence of spectra. Hempel and Voigt [39] proved that the spectrum is p-independent for a large class of Schrödinger operators acting on $L^p(\mathbb{R}^n)$. Arendt [3] extended this to elliptic operators on domains of $\mathbb{R}^n$ under the assumption that the semi-group of the given operator satisfies $L^1 - L^\infty$ off-diagonal estimates. Ouhabaz realized in his PhD thesis (published later in [58]) that the same assumption yields holomorphy of the semi-group up to L^1 with optimal angles in the case of a self-adjoint operator on a subset of $\mathbb{R}^n$. Independently, Arendt & ter Elst [4, Theorem 5.4], and Hieber [40] removed the self-adjointness assumption. Finally, Davies [31] extended Ouhabaz argument to the setting of doubling spaces (for non-negative self-adjoint operators) to obtain in a simpler manner the result of p-independence of spectrum of Arendt.

3.6. ANALYTIC EXTENSION

Recall that $\mathcal{J}(L)$ (resp. $\mathcal{N}(L)$) is the maximal interval of exponents p in $[1, \infty]$ for which the semigroup $(e^{-tL})_{t>0}$ (resp. the family $(\sqrt{t}\,\nabla e^{-tL})_{t>0}$) is L^p bounded.

PROPOSITION 3.15. *Let ω be the type of L. Then the semigroup $(e^{-tL})_{t>0}$ has an analytic extension to $\Sigma_{\frac{\pi}{2}-\omega}$ on L^p for $p \in \operatorname{Int} \mathcal{J}(L)$. Moreover, for $p \in \operatorname{Int} \mathcal{J}(L)$ and all $\beta \in (0, \frac{\pi}{2}-\omega)$, the families $(e^{-zL})_{z\in\Sigma_\beta}$ and $(zLe^{-zL})_{z\in\Sigma_\beta}$ are L^p bounded, satisfy $L^p - L^2$ (resp. $L^2 - L^p$) off-diagonal estimates and are $L^p - L^2$ (resp. $L^2 - L^p$) bounded if $p < 2$ (resp. if $p > 2$).*

PROOF. The analyticity is a consequence of Stein's complex interpolation theorem together with holomorphy of the semigroup on L^2. However, this does not yield the best angle for the sector of holomorphy. Here is a better argument.

For $\alpha \in (-\frac{\pi}{2}+\omega, \frac{\pi}{2}-\omega)$, set $L_\alpha = e^{i\alpha}L$. It is an operator in the same class as L, associated to the matrix of coefficients $e^{i\alpha}A(x)$. Hence, Proposition 3.2 applies to the semigroup associated to L_α. Furthermore, a careful check of the proof shows that the various constants are independent of α as long as α is restricted to a compact subset of $(-\frac{\pi}{2}+\omega, \frac{\pi}{2}-\omega)$. Let $\beta \in (0, \frac{\pi}{2}-\omega)$. If $z = te^{i\alpha} \in \Sigma_\beta$, then $e^{-zL} = e^{-tL_\alpha}$ and the reasonning above shows that the statement of Proposition 3.2 extends to the complex family $(e^{-zL})_{z\in\Sigma_\beta}$. Hence, for $p \in \operatorname{Int} \mathcal{J}(L)$, it remains to showing that this family is $L^p - L^2$ (resp. $L^2 - L^p$) bounded if $p < 2$ (resp. if $p > 2$).

The case $p > 2$ can be handled by duality, we restrict attention to $p < 2$. Let $z \in \Sigma_\beta$. Choose $\beta < \beta' < \frac{\pi}{2} - \omega$. Elementary geometry shows that one can decompose $z = \zeta + t$ with $\zeta \in \Sigma_{\beta'}$, $t > 0$ and $|z| \sim |\zeta| \sim t$ where the implicit constants only depend on β and β'. By assumption and Proposition 3.2, $(e^{-tL})_{t>0}$ is $L^p - L^2$ bounded. Since $(e^{-\zeta L})_{\zeta\in\Sigma_{\beta'}}$ is L^2 bounded, writing $e^{-zL} = e^{-\zeta L}e^{-tL}$, this shows that $(e^{-zL})_{z\in\Sigma_\beta}$ is $L^p - L^2$ bounded. $\square$

PROPOSITION 3.16. *Let ω be the type of L. Then, for $p \in \operatorname{Int} \mathcal{N}(L)$ and all $\beta \in (0, \frac{\pi}{2}-\omega)$, its analytic extension $(\sqrt{z}\,\nabla e^{-zL})_{z\in\Sigma_\beta}$ is L^p bounded, satisfies $L^p - L^2$ (resp. $L^2 - L^p$) off-diagonal estimates and is $L^p - L^2$ (resp. $L^2 - L^p$) bounded if $p < 2$ (resp. if $p > 2$).*

It suffices to adapt the above proof. Further details are left to the reader.

CHAPTER 4

L^p theory for square roots

Let L be as the introduction. We study here the following sets:
(1) the maximal interval of exponents p in $(1, \infty)$ for which one has the L^p boundedness of the Riesz transform $\nabla L^{-1/2}$, which we call $\mathcal{I}(L)$.
(2) the maximal interval of exponents p in $(1, \infty)$ for which one has the *a priori* inequality $\|L^{1/2}f\|_p \lesssim \|\nabla f\|_p$ for $f \in C_0^\infty$, which we call $\mathcal{R}(L)$.

We characterize $\mathcal{I}(L)$ and obtain bounds on the limits of $\mathcal{R}(L)$.

4.1. Riesz transforms on L^p

We prove here the following theorem.

THEOREM 4.1. *The interior of $\mathcal{I}(L)$ equals $(p_-(L), q_+(L))$.*

Recall that $p_-(L)$ is the lower limit of both $\mathcal{J}(L)$ and $\mathcal{N}(L)$, and that $q_+(L)$ is the upper limit of $\mathcal{N}(L)$. The cases $p < 2$ and $p > 2$ are treated by different methods. See Section 4.3 for an improvement of this result.

4.1.1. The case $p < 2$. We introduce the following sets.

$\mathcal{I}_-(L) = \mathcal{I}(L) \cap (1, 2)$

$\mathcal{J}_-(L) = \mathcal{J}(L) \cap [1, 2)$

$\mathcal{K}_-(L) = \{1 \le p < 2; (e^{-tL})_{t>0} \text{ is } L^p - L^2 \text{ bounded}\}$

$\mathcal{M}_-(L) = \{1 \le p < 2; (e^{-tL})_{t>0} \text{ satisfies } L^p - L^2 \text{ off} - \text{diagonal estimates}\}$

It is quite clear from interpolation that these sets are intervals with 2 as upper limit, provided they are nonempty. It follows directly from Proposition 3.2 that $\mathcal{J}_-(L)$, $\mathcal{K}_-(L)$ and $\mathcal{M}_-(L)$ have the same interiors (that is, the same lower limit). Also, by Corollary 3.5 the interior of $\mathcal{J}_-(L)$ is not empty.

THEOREM 4.2. [1] *All the above sets are intervals with common interiors.*

Let us first derive a corollary for results in the range $1 < p < 2$.

PROPOSITION 4.3. *$\nabla L^{-1/2}$ is bounded on L^p for $1 < p < 2$ if, and only if, $(e^{-tL})_{t>0}$ is $L^p - L^2$ bounded for $1 < p < 2$. In particular, if $(e^{-tL})_{t>0}$ is $L^1 - L^2$ bounded, then $\nabla L^{-1/2}$ is bounded on L^p for $1 < p < 2$.*

PROOF. The equivalence is contained in the theorem above since $\operatorname{Int} \mathcal{I}_-(L) = \operatorname{Int} \mathcal{K}_-(L)$. Next, if $(e^{-tL})_{t>0}$ is $L^1 - L^2$ bounded, then by interpolation it is also $L^p - L^2$ bounded for $1 < p < 2$ □

---

[1] The fact that off-diagonal estimates implies boundedness of the Riesz transforms is proved in [21] and this idea is also the main tool in [42]. The converse was not noticed in these works.

REMARK. Usually, the L^p boundedness of the Riesz transform for $1 < p < 2$ is obtained under the stronger assumption that the semigroup kernel has good pointwise upper bounds.[2]

PROOF OF THEOREM 4.2. According to the above remark, the following steps suffice to prove this result:

Step 1: $p \in \mathcal{I}_-(L)$ implies $p \in \mathcal{K}_-(L)$.
Step 2: $p_0 \in \mathcal{M}_-(L)$ and $p_0 < p < 2$ imply $p \in \mathcal{I}_-(L)$.

Proof of Step 1: $p \in \mathcal{I}_-(L)$ implies $p \in \mathcal{K}_-(L)$. First the conclusion is true if $n \leq 2$ from Corollary 3.6. We assume, therefore, that $n \geq 3$.

Let $p \in \mathcal{I}_-(L)$. If $p \geq 2_* = \frac{2n}{n+2}$, we already know that $(e^{-tL})_{t>0}$ is $L^p - L^2$. Assume now that $p < 2_*$. Remark that any $q \in [p, 2]$ belongs to $\mathcal{I}_-(L)$. By Sobolev embedding's, we see that $L^{-1/2} \colon L^q \to L^{q^*}$. Let $p_0 = p$ and $p_k = (p_{k-1})^*$ and stop when k is the largest integer that satisfies $p_k < 2$. We have that $L^{-k/2}$ is bounded from L^p into L^{p_k}. Write for $f \in L^2 \cap L^p$,
$$e^{-tL} f = (e^{-(t/2)L} L^{k/2}) e^{-(t/2)L} L^{-k/2} f,$$
the equality being justified by the fact that $f \in L^2$. We successively have that $g = L^{-k/2} f \in L^{p_k}$, then $h = e^{-(t/2)L} g \in L^2$ by Corollary 3.5 since $p_k \geq \frac{2n}{n+2}$, and $e^{-(t/2)L} L^{k/2} h \in L^2$ by the bounded holomorphic calculus of L on L^2. Hence, e^{-tL} is bounded from L^p into L^2 by density and we obtain the right bound for its operator norm by keeping track of the bounds from each step.

Proof of Step 2: $p_0 \in \mathcal{M}_-(L)$ and $p_0 < p < 2$ imply $p \in \mathcal{I}_-(L)$. We apply Theorem 1.1 to the operator $T = \nabla L^{-1/2}$ to obtain weak type (p_0, p_0). We first introduce the operators $A_r = I - (I - e^{-r^2 L})^m$ where m is some integer to be specified later.

Observe that $A_r = \sum_{k=1}^m c_k e^{-kr^2 L}$ for some numbers c_k. A direct consequence of the assumption that L satisfies $L^{p_0} - L^2$ off-diagonal estimates is that if B is any ball, with radius r, and $f \in L^2 \cap L^{p_0}$ with support in B, and $j \geq 1$, then

$$\left(\fint_{C_j(B)} |A_r f|^2 \right)^{1/2} \leq C r^{-\gamma} e^{-\alpha 4^j} \left(\fint_B |f|^{p_0} \right)^{1/p_0}$$

where $\gamma = \gamma_{p_0} = |\frac{n}{2} - \frac{n}{p_0}|$. Recall that $C_j(B)$ denotes the ring $2^{j+1} B \setminus 2^j B$ if $j \geq 2$ and $C_1(B) = 4B$. Hence the assumption (1.2) in Theorem 1.1 holds with $g(j) = C(m) 2^{-nj/2} e^{-\alpha 4^j}$ for any $m \geq 1$. It remains to check the assumption (1.1) and this is where we use the role of m.

LEMMA 4.4. Let $p_0 \in \mathcal{M}_-(L)$. There exists $C \geq 0$, such that for all balls B with radius $r > 0$ and $f \in L^2 \cap L^{p_0}$ with support in B, and $j \geq 2$

(4.1) $\quad \|\nabla L^{-1/2} (I - e^{-r^2 L})^m f\|_{L^2(C_j(B))} \leq C r^{-\gamma} 2^{-j(2m+\gamma)} \|f\|_{L^{p_0}(B)},$

where $\gamma = \gamma_{p_0} = |\frac{n}{2} - \frac{n}{p_0}|$.

---

[2] In [**17**, Chapter IV], this is obtained via an $\mathcal{H}^1$ estimate under $L^1 - L^\infty$ off diagonal estimates plus Hölder regularity on the heat kernel. The regularity was removed in [**34**] to obtain weak type (1,1). In [**27**], the use of weighted L^2 estimates similar to the $L^1 - L^2$ boundedness was stressed in the context of the Riesz transform for the Laplace-Beltrami on Riemannian manifolds.

PROOF. By expanding $(I - e^{-r^2 L})^m$ in the representation of the square root, we obtain

$$\begin{aligned}
\nabla L^{-1/2}(I - e^{-r^2 L})^m f &= \pi^{-1/2} \int_0^\infty \nabla e^{-tL}(I - e^{-r^2 L})^m f \, \frac{dt}{\sqrt{t}} \\
&= \pi^{-1/2} \int_0^\infty g_{r^2}(t) \nabla e^{-tL} f \, dt
\end{aligned}$$

where using the usual notation for the binomial coefficient,

$$g_s(t) = \sum_{k=0}^m \binom{m}{k} (-1)^k \frac{\chi(t - ks)}{\sqrt{t - ks}},$$

and χ is the indicator function of $(0, \infty)$.

Observe that $(\sqrt{t} \nabla e^{-tL})_{t>0}$ satisfies $L^{p_0} - L^2$ off-diagonal estimates by applying the composition to the families $(\sqrt{t} \nabla e^{-(t/2)L})_{t>0}$ and $(e^{-(t/2)L})_{t>0}$ which satisfy respectively L^2 and $L^{p_0} - L^2$ off-diagonal estimates. By Minkowski integral inequality, we have that

$$\|\nabla L^{-1/2}(I - e^{-r^2 L})^m f\|_{L^2(C_j(B))} \leq C \int_0^\infty |g_{r^2}(t)| e^{-\frac{\alpha 4^j r^2}{t}} t^{-\gamma/2} \frac{dt}{\sqrt{t}} \|f\|_{L^{p_0}(B)}.$$

The latter integral can be estimated as follows. Elementary analysis yields the following estimates for g_{r^2}:

$$|g_{r^2}(t)| \leq \frac{C}{\sqrt{t - kr^2}} \quad \text{if} \quad kr^2 < t \leq (k+1)r^2 \leq (m+1)r^2$$

and

$$|g_{r^2}(t)| \leq C r^{2m} t^{-m - \frac{1}{2}} \quad \text{if} \quad t > (m+1)r^2.$$

The latter estimate comes from the inequality

$$\left| \sum_{k=0}^m \binom{m}{k} (-1)^k v(t - ks) \right| \leq C s^m \sup_{u \geq t/(m+1)} |v^{(m)}(u)|$$

for $t > (m+1)s \geq 0$ after expanding $v(t - ks)$ using Taylor's formula about t and using the classical relations $\sum_{k=0}^m \binom{m}{k}(-1)^k k^\ell = 0$ for $\ell \in \mathbb{N}$, $\ell < m$. This readily yields the estimates

$$\int_0^\infty |g_{r^2}(t)| e^{-\frac{\alpha 4^j r^2}{t}} t^{-\gamma/2} \frac{dt}{\sqrt{t}} \leq C r^{-\gamma} 2^{-j(2m+\gamma)}.$$

Alternate proof of the lemma: By the representation of the square root, one obtains

$$(4.2) \qquad \nabla L^{-1/2}(I - e^{-r^2 L})^m f = \pi^{-1/2} \int_0^\infty \sqrt{t} \, \nabla e^{-tL}(I - e^{-r^2 L})^m f \, \frac{dt}{t}.$$

The function $\varphi(z) = e^{-tz}(1 - e^{r^2 z})^m$ is holomorphic and satisfies the technical condition (2.5) in any sector Σ_μ, $\mu < \frac{\pi}{2}$. Hence, one may use the representation (2.6) to compute $\varphi(L)$. With the same choices of the parameters as in Section 2.2, for some positive constant c, the functions $\eta_\pm$ given by (2.7) satisfy

$$|\eta_\pm(z)| \leq \int_{\gamma_\pm} e^{-c|\zeta|(|z|+t)} |1 - e^{-r^2 \zeta}|^m \, |d\zeta|,$$

and one finds

(4.3) $$|\eta_\pm(z)| \leq \frac{C}{|z|+t} \inf\left(1, \frac{r^{2m}}{(|z|+t)^m}\right), \quad z \in \Gamma_\pm.$$

Observe that for any $0 < \beta < \beta' < \frac{\pi}{2} - \omega$, the family $(\sqrt{z}\,\nabla e^{-zL})_{z \in \Sigma_\beta}$ satisfies $L^{p_0} - L^2$ off-diagonal estimates by applying the composition lemma to the families $(\sqrt{z}\,\nabla e^{-zL})_{z \in \Sigma_{\beta'}}$ and $(e^{-tL})_{t>0}$ which satisfy respectively L^2 and $L^{p_0} - L^2$ off-diagonal estimates. It suffices to decompose $z \in \Sigma_\beta$ into $z' + t$ with $z' \in \Sigma_{\beta'}$ and $t > 0$ with $|z| \sim |z'| \sim t$. Using this in (2.6) and the estimate for $\eta_\pm$, $\|\sqrt{t}\,\nabla e^{-tL}(I - e^{-r^2 L})^m f\|_{L^2(C_j(B))}$ is bounded by

$$C \int_{\Gamma_+} e^{-\frac{c 4^j r^2}{|z|}} \frac{t^{1/2}}{|z|^{1/2}} \frac{1}{|z|^{\gamma/2}} \frac{1}{(|z|+t)} \frac{r^{2m}}{(|z|+t)^m} |dz|\, \|f\|_{L^{p_0}(B)}$$

plus the similar term corresponding to integration on Γ_-. Here $\gamma = \gamma_{p_0} = |\frac{n}{2} - \frac{n}{p_0}|$. Using the inequality (4.4) below for the integral, this gives us the bound

$$\frac{C}{4^{jm}(2^j r)^\gamma} \inf\left(\left(\frac{t}{4^j r^2}\right)^{1/2}, \left(\frac{4^j r^2}{t}\right)^{m-1/2}\right) \|f\|_{L^{p_0}(B)}.$$

Integrating with respect to t in (4.2), one finds (4.1) by Minkowski inequality since $m \geq 1 > 1/2$. $\square$

Thus, (1.1) holds with $g(j) = 2^{-nj/2} 2^{-j(2m+\gamma)}$. Hence, the summability condition $\sum_{j \geq 1} 2^{nj} g(j) < \infty$ is granted if $2m + \gamma > n/2$. This finishes the proof of Step 2 modulo the proof of (4.4) which we do in the following lemma. $\square$

LEMMA 4.5. *Let $\gamma \geq 0$, $\alpha \geq 0$, $m > 0$ be fixed parameters, and c a positive constant. For some C independent of $j \in \mathbb{N}$, $r > 0$ and $t > 0$, the integral*

$$I = \int_0^\infty e^{-\frac{c 4^j r^2}{s}} \frac{1}{s^{\gamma/2}} \frac{t^\alpha}{(s+t)^{1+\alpha}} \frac{r^{2m}}{(s+t)^m}\, ds$$

satisfies the estimate

(4.4) $$I \leq \frac{C}{4^{jm}(2^j r)^\gamma} \inf\left(\left(\frac{t}{4^j r^2}\right)^\alpha, \left(\frac{4^j r^2}{t}\right)^m\right).$$

PROOF. Set $\beta = \frac{4^j r^2}{t}$ for the argument. Split $I = I_1 + I_2$ where $I_1 = \int_t^\infty \ldots$ and $I_2 = \int_0^t \ldots$.

If $t \leq s$, write $\frac{t^\alpha}{(s+t)^{1+\alpha}} \leq \frac{t^\alpha}{s^{1+\alpha}}$, $\frac{r^{2m}}{(s+t)^m} \leq \frac{r^{2m}}{s^m}$ and change variable by setting $\frac{4^j r^2}{s} = u$. Then,

$$I_1 \leq \int_0^{\frac{4^j r^2}{t}} e^{-cu} \left(\frac{u}{4^j r^2}\right)^{\gamma/2} \left(\frac{tu}{4^j r^2}\right)^\alpha \frac{u^m}{4^{jm}} \frac{du}{u} = \frac{\beta^{-\alpha}}{4^{jm}(2^j r)^\gamma} \int_0^\beta e^{-cu} u^{\gamma/2 + \alpha + m} \frac{du}{u}.$$

As $\gamma/2 + \alpha + m > 0$, we obtain

$$I_1 \leq \frac{C}{4^{jm}(2^j r)^\gamma} \inf\left(\beta^{\gamma/2 + m}, \beta^{-\alpha}\right).$$

If $t \geq s$ and $\gamma > 0$, write $\frac{t^\alpha}{(s+t)^{1+\alpha}} \leq \frac{1}{s}$, $\frac{r^{2m}}{(s+t)^m} \leq \frac{r^{2m}}{t^m}$ and change variable by setting $\frac{4^j r^2}{s} = u$. Then,

$$I_2 \leq \int_{\frac{4^j r^2}{t}}^\infty e^{-cu} \left(\frac{u}{4^j r^2}\right)^{\gamma/2} \frac{r^{2m}}{t^m} \frac{du}{u} = \frac{\beta^m}{4^{jm}(2^j r)^\gamma} \int_\beta^\infty e^{-cu} u^{\gamma/2} \frac{du}{u}.$$

Since $\gamma > 0$, we obtain

$$I_2 \leq \frac{C}{4^{jm}(2^j r)^\gamma} \inf\left(\beta^m, e^{-c\beta}\right)$$

where the value of c has changed. The conclusion follows readily.

To treat the case $\gamma = 0$ when $\beta \leq 1$, then use instead $\frac{t^\alpha}{(s+t)^{1+\alpha}} \leq \frac{1}{t}$ and that $\int_\varepsilon^\infty e^{-cu} \frac{du}{u^2}$ is asymptotic to $c\varepsilon^{-1}$ as ε tends to 0. Further details are left to the reader. $\square$

4.1.2. The case $p > 2$. Introduce the following sets:

$\mathcal{I}_+(L) = \mathcal{I}(L) \cap (2, \infty)$

$\mathcal{N}_+(L) = \mathcal{N}(L) \cap (2, \infty]$

$\mathcal{K}_+(L) = \{2 < p \leq \infty; (\sqrt{t}\nabla e^{-tL})_{t>0} \text{ is } L^2 - L^p \text{ bounded}\}$

$\mathcal{M}_+(L) = \{2 < p \leq \infty; (\sqrt{t}\nabla e^{-tL})_{t>0} \text{ satisfies } L^2 - L^p \text{ off} - \text{diagonal estimates}\}$

Observe that these sets, if nonempty, are intervals with 2 as lower limit. We know that $\mathcal{N}_+(L)$ is not empty and that it has the same interior as $\mathcal{K}_+(L)$ and $\mathcal{M}_+(L)$.

THEOREM 4.6. *The above sets are intervals with common interiors.*

Let us first state a consequence for the range $2 < p < \infty$.

PROPOSITION 4.7. *$\nabla L^{-1/2}$ is bounded on L^p for $2 < p < \infty$ if, and only if, $(\sqrt{t}\nabla e^{-tL})_{t>0}$ is $L^2 - L^p$ bounded for $2 < p < \infty$. In particular, if $(\sqrt{t}\nabla e^{-tL})_{t>0}$ is $L^2 - L^\infty$ bounded, then $\nabla L^{-1/2}$ is bounded on L^p for $2 < p < \infty$.*

PROOF. The equivalence is contained in the theorem above. If $(\sqrt{t}\nabla e^{-tL})_{t>0}$ is $L^2 - L^\infty$ bounded then by interpolation with the L^2-boundedness, it is also $L^2 - L^p$ bounded for $2 < p < \infty$. $\square$

PROOF OF THEOREM 4.6. In view of the above remarks, it is enough to show

Step 1: $p \in \mathcal{I}_+(L)$ implies $p \in \mathcal{K}_+(L)$.
Step 2: $p_0 \in \mathcal{M}_+(L)$ and $2 < p < p_0$ imply $p \in \mathcal{I}_+(L)$.

Proof of Step 1: $p \in \mathcal{I}_+(L)$ implies $p \in \mathcal{K}_+(L)$. Assume first that $n \geq 3$. By Sobolev embedding's, if $q \in [2, p]$ with $q < n$ then

$$L^{-1/2} \colon L^q \to L^{q^*}.$$

Following the method of Subsection 4.1.1 applied to the dual semigroup, we obtain that if $q \in \mathbb{R}$ with $2 \leq q \leq p^*$,

$$e^{-tL} \colon L^2 \to L^q$$

with bound $Ct^{-\gamma_q}/2$. Since, $L^{1/2}e^{-(t/2)L}$ is bounded on L^2 with bound $Ct^{-1/2}$, we obtain $\nabla e^{-tL} = \nabla L^{-1/2}e^{-(t/2)L}L^{1/2}e^{-(t/2)L}$ is bounded from L^2 to L^p with the desired bound on the operator norm.

If $n \leq 2$, we already know that
$$e^{-tL} \colon L^2 \to L^q$$
for all $q \in [2, \infty]$ (See Chapter 3). The argument is then similar. This ends the proof of Step 1.

Proof of Step 2: $p_0 \in \mathcal{M}_+(L)$ and $2 < p < p_0$ imply $p \in \mathcal{I}_+(L)$. We apply Theorem 1.2 to $T = \nabla L^{-1/2}$. We set again $A_r = I - (I - e^{-r^2 L})^m$ for some integral number m to be chosen. We have to check (1.3) and (1.4).

We begin with the following lemma.

LEMMA 4.8. *For every ball B with radius $r > 0$,*

$$(4.5) \quad \|\nabla L^{-1/2}(I - e^{-r^2 L})^m f\|_{L^2(B)} \leq |B|^{1/2} \sum_{j \geq 1} g(j) \left(\frac{1}{|2^{j+1}B|} \int_{2^{j+1}B} |f|^2 \right)^{1/2}$$

with $g(j) = C 2^{(n/2)j} 4^{-mj}$.

Hence, (1.3) follows provided $m > n/4$, since $\left(\frac{1}{|2^{j+1}B|} \int_{2^{j+1}B} |f|^2 \right)^{1/2}$ is controlled by $\left(M(|f|^2)(y) \right)^{1/2}$ for any $y \in B$.

PROOF. Fix a ball B and $r = r(B)$ its radius. Decompose f as $f_1 + f_2 + f_3 + \ldots$ with $f_j = f\chi_{C_j}$ where χ_{C_j} is the indicator function of $C_j = C_j(B)$. By Minkowski inequality we have that

$$\|\nabla L^{-1/2}(I - e^{-r^2 L})^m f\|_{L^2(B)} \leq \sum_{j \geq 1} \|\nabla L^{-1/2}(I - e^{-r^2 L})^m f_j\|_{L^2(B)}.$$

For $j = 1$ we use L^2 boundedness of $\nabla L^{-1/2}(I - e^{-r^2 L})^m$:

$$\|\nabla L^{-1/2}(I - e^{-r^2 L})^m f_1\|_{L^2(B)} \lesssim \|f\|_{L^2(4B)} \leq |4B|^{1/2} \left(\frac{1}{|4B|} \int_{4B} |f|^2 \right)^{1/2}.$$

First proof in the case $j \geq 2$: Expanding $(I - e^{-r^2 L})^m$ in the representation of the square root, we find

$$\nabla L^{-1/2}(I - e^{-r^2 L})^m f_j = \pi^{-1/2} \int_0^\infty \nabla e^{-tL}(I - e^{-r^2 L})^m f_j \frac{dt}{\sqrt{t}}$$
$$= \pi^{-1/2} \int_0^\infty g_{r^2}(t) \nabla e^{-tL} f_j \, dt$$

where, as in Subsection 4.1.1,

$$g_s(t) = \sum_{k=0}^m \binom{m}{k} (-1)^k \frac{\chi(t - ks)}{\sqrt{t - ks}}.$$

By Minkowski integral inequality and the L^2 off-diagonal estimates for the family $(\sqrt{t}\nabla e^{-tL})_{t>0}$ using the support of f_j, we have that

$$\|\nabla L^{-1/2}(I - e^{-r^2 L})^m f_j\|_{L^2(B)} \leq C \int_0^\infty |g_{r^2}(t)| e^{-\frac{\alpha 4^j r^2}{t}} \frac{dt}{\sqrt{t}} \|f\|_{L^2(C_j)}.$$

As in Subsection 4.1.1, the latter integral is bounded above by $C4^{-jm}$ uniformly over $r > 0$. Next,

$$\|f\|_{L^2(C_j)} \leq |2^{j+1}B|^{1/2} \left(\frac{1}{|2^{j+1}B|} \int_{2^{j+1}B} |f|^2\right)^{1/2}$$

and we obtain (4.5).

Second proof in the case $j \geq 2$: Write again (4.2):

$$\nabla L^{-1/2}(I - e^{-r^2 L})^m f_j = \pi^{-1/2} \int_0^\infty \sqrt{t}\, \nabla e^{-tL}(I - e^{-r^2 L})^m f_j \, \frac{dt}{t}.$$

As in Subsection 4.1.1, one may use the representation (2.6) with the function $\varphi(z) = e^{-tz}(1 - e^{r^2 z})^m$. The functions $\eta_\pm$ in (2.7) satisfy the estimates (4.3). Since for any $0 < \beta < \frac{\pi}{2} - \omega$, $(\sqrt{z}\,\nabla e^{-zL})_{z \in \Sigma_\beta}$ satisfies L^2 off-diagonal estimates, using (2.6) and the estimate for $\eta_\pm$, $\|\sqrt{t}\,\nabla e^{-tL}(I - e^{-r^2 L})^m f_j\|_{L^2(B)}$ is bounded by

$$C \int_{\Gamma_+} e^{-\frac{c 4^j r^2}{|z|}} \frac{t^{1/2}}{|z|^{1/2}} \frac{1}{(|z|+t)} \frac{r^{2m}}{(|z|+t)^m} |dz| \, \|f\|_{L^2(C_j)}$$

plus the similar term corresponding to integration on Γ_-. Using (4.4), this gives us the bound

$$\frac{C}{4^{jm}} \inf\left(\left(\frac{t}{4^j r^2}\right)^{1/2}, \left(\frac{4^j r^2}{t}\right)^{m-1/2}\right) \|f\|_{L^2(C_j)}.$$

Integrating with respect to t in (4.2), one finds (4.5) by Minkowski inequality since $m \geq 1 > 1/2$. $\square$

We now show that (1.4) holds. By expanding $A_r = I - (I - e^{-r^2 L})^m$ it suffices to show

(4.6) $$\left(\frac{1}{|B|}\int_B |\nabla e^{-kr^2 L} f|^{p_0}\right)^{1/p_0} \leq \sum_{j \geq 1} g(j) \left(\frac{1}{|2^{j+1}B|}\int_{2^{j+1}B} |\nabla f|^2\right)^{1/2}$$

for every ball B with $r = r(B)$, and $k = 1, 2, \ldots, m$ with $\sum g(j) < \infty$. Recall that m is chosen larger than $n/4$. This, applied to $f = L^{-1/2}g$ for appropriate g, gives us (1.4).

Using the conservation property (2.12) of the semigroup, we have

$$\nabla e^{-kr^2 L} f = \nabla e^{-kr^2 L}(f - f_{4B})$$

where f_E is the mean of f over E. Write $f - f_{4B} = f_1 + f_2 + f_3 + \ldots$ where $f_j = (f - f_{4B})\chi_{C_j(B)}$. For $j = 1$, we use the fact that $p_0 \in \mathcal{K}_+$, that is that $(\sqrt{t}\,\nabla e^{-tL})_{t>0}$ satisfies $L^2 - L^{p_0}$ off-diagonal estimates, and Poincaré inequalities to obtain

$$\left(\frac{1}{|B|}\int_B |\nabla e^{-kr^2 L} f_1|^{p_0}\right)^{1/p_0} \leq C \left(\frac{1}{r^2|4B|}\int_{4B} |f - f_{4B}|^2\right)^{1/2}$$

$$\leq C \left(\frac{1}{|4B|}\int_{4B} |\nabla f|^2\right)^{1/2}$$

For $j \geq 2$, we have similarly by the assumption on p_0 and the various support assumptions,

$$\left(\frac{1}{|B|}\int_B |\nabla e^{-kr^2 L} f_j|^{p_0}\right)^{1/p_0} \leq \frac{Ce^{-\alpha 4^j} 2^{jn/2}}{r}\left(\frac{1}{|2^{j+1}B|}\int_{C_j(B)} |f_j|^2\right)^{1/2}.$$

But

$$\int_{C_j(B)} |f_j|^2 \leq \int_{2^{j+1}B} |f - f_{4B}|^2,$$

$$|f - f_{4B}| \leq |f - f_{2^{j+1}B}| + \sum_{\ell=2}^{j} |f_{2^\ell B} - f_{2^{\ell+1}B}|$$

and observe that by Poincaré inequality,

$$|f_{2^\ell B} - f_{2^{\ell+1}B}|^2 \leq \frac{2^n}{|2^{\ell+1}B|}\int_{2^{\ell+1}B} |f - f_{2^{\ell+1}B}|^2 \leq \frac{C(2^\ell r)^2}{|2^{\ell+1}B|}\int_{2^{\ell+1}B} |\nabla f|^2.$$

Hence, by Minkowski inequality, we easily obtain

$$\left(\frac{1}{|B|}\int_B |\nabla e^{-kr^2 L} f_j|^{p_0}\right)^{1/p_0} \leq Ce^{-\alpha 4^j} 2^{jn/2} \sum_{\ell=1}^{j} 2^\ell \left(\frac{1}{|2^{\ell+1}B|}\int_{2^{\ell+1}B} |\nabla f|^2\right)^{1/2}$$

and summing over $j \geq 2$ gives us (4.6).

This concludes the proof of Step 2, hence that of Theorem 4.6. □

4.2. Reverse inequalities

In this section, we study the reverse inequality to the Riesz transform L^p boundedness. Recall the maximal interval, $\mathcal{R}(L)$, of exponents p in $(1,\infty)$ for which one has the *a priori* inequality $\|L^{1/2} f\|_p \lesssim \|\nabla f\|_p$ for $f \in C_0^\infty$ and write $\operatorname{Int}\mathcal{R}(L) = (r_-(L), r_+(L))$. We know so far that $\operatorname{Int}\mathcal{I}(L) = (p_-(L), q_+(L))$ and that $p_+(L) \geq (q_+(L))^*$. We show the following bounds.

THEOREM 4.9. *We have*

$$r_-(L) \leq \sup(1, (p_-(L))_*),$$
$$r_+(L) \geq p_+(L).$$

Hence, $\mathcal{R}(L)$ contains a neighborhood of the closure (in $\mathbb{R}$) of $\mathcal{I}(L)$.

We discuss the optimality of these bounds in Section 5.4.

This result is a consequence of Lemma 4.10 and Proposition 4.11 below. We begin with a duality principle which applies for all p in $(1,\infty)$ but which gives us the bound in Theorem 4.9 only for $r_+(L)$.

LEMMA 4.10. *If $1 < p < \infty$ and $\nabla L^{-1/2}$ is bounded on L^p then $\|(L^*)^{1/2} f\|_{p'} \lesssim \|\nabla f\|_{p'}$ holds for $f \in C_0^\infty$. Hence, $r_+(L) \geq p_+(L)$.*

PROOF. Let $f \in C_0^\infty$. Then $(L^*)^{1/2} f$ is defined and belongs to L^2. We estimate $\|(L^*)^{1/2} f\|_{p'}$ by testing against $g \in L^2 \cap L^p$. Since $g \in L^2$, we have $g = L^{1/2} L^{-1/2} g$ and $h = L^{-1/2} g \in \dot{W}^{1,2}$. Hence, by Lemma 2.2,

$$\langle (L^*)^{1/2} f, g \rangle = \int_{\mathbb{R}^n} \nabla f \cdot \overline{A \nabla h}.$$

Since $\|\nabla h\|_p \lesssim \|g\|_p$, it follows that $\|(L^*)^{1/2} f\|_{p'} \lesssim \|\nabla f\|_{p'}$ as desired.

Let $2 < p < p_+(L)$. By duality, $p_-(L^*) < p' < 2$, hence, $\nabla(L^*)^{-1/2}$ is bounded on $L^{p'}$ by Theorem 4.2 and $p \in \mathcal{R}(L)$. Thus, $r_+(L) > p$ and the conclusion follows. $\square$

The above lemma does not give an interesting information for $r_-(L)$. The next proposition yields a much better bound. Define $\tilde{p}_-(L) = \sup(1, (p_-(L))_*)$.

PROPOSITION 4.11. *Let $1 < p < 2$. If $\nabla L^{-1/2}$ is bounded on L^p and $p \geq \frac{n}{n-1}$ then $\|L^{1/2}f\|_q \lesssim \|\nabla f\|_q$ for $p_* < q < 2$. If $\nabla L^{-1/2}$ is bounded on L^p and $p < \frac{n}{n-1}$ then $\|L^{1/2}f\|_q \lesssim \|\nabla f\|_q$ for $1 < q < 2$. Hence, $r_-(L) \leq \tilde{p}_-(L)$.*

The inequality on $r_-(L)$ follows from the definition of $r_-(L)$ and the identification of $p_-(L)$ as the lower limit of $\mathcal{I}(L)$.

Let us remark that for $n \geq 2$ and $p < 2$, there exists an L such that $p \in \mathcal{I}_-(L)$ but $p' \notin \mathcal{I}_+(L^*)$. [3] Hence, no duality argument can help us here. In fact, the argument will rely on a Calderón-Zygmund decomposition for Sobolev functions which is proved in the Appendix.

LEMMA 4.12. *Let $n \geq 1$, $1 \leq p \leq \infty$ and $f \in \mathcal{D}'(\mathbb{R}^n)$ be such that $\|\nabla f\|_p < \infty$. Let $\alpha > 0$. Then, one can find a collection of cubes (Q_i), functions g and b_i such that*

$$(4.7) \qquad f = g + \sum_i b_i$$

and the following properties hold:

$$(4.8) \qquad \|\nabla g\|_\infty \leq C\alpha,$$

$$(4.9) \qquad b_i \in W_0^{1,p}(Q_i) \text{ and } \int_{Q_i} |\nabla b_i|^p \leq C\alpha^p |Q_i|,$$

$$(4.10) \qquad \sum_i |Q_i| \leq C\alpha^{-p} \int_{\mathbb{R}^n} |\nabla f|^p,$$

$$(4.11) \qquad \sum_i \mathbf{1}_{Q_i} \leq N,$$

where C and N depend only on dimension and p.

The space $W_0^{1,p}(\Omega)$ denotes the closure of $C_0^\infty(\Omega)$ in $W^{1,p}(\Omega)$. The point is in the fact that the functions b_i are supported in cubes as the original Calderón-Zygmund decomposition applied to ∇f would not give this.

PROOF OF PROPOSITION 4.11. By Theorem 4.2 we may transform the hypothesis on the Riesz transform, i.e. $p \in \mathcal{I}_-(L)$, into an hypothesis on the semigroup. Proposition 4.11 is therefore a consequence of the next result combined with Marcinkiewicz interpolation. $\square$

LEMMA 4.13. *Let $\rho \in \mathcal{M}_-(L)$. Then we have*

$$(4.12) \qquad \|L^{1/2}f\|_{p,\infty} \lesssim \|\nabla f\|_p, \quad \text{if } 1 \leq \rho_* < p < 2,$$

$$(4.13) \qquad \|L^{1/2}f\|_{1,\infty} \lesssim \|\nabla f\|_1, \quad \text{if } \rho_* < 1.$$

---

[3] This is due to Kenig. See [**17**, Chapter IV, Theorem 7].

PROOF. Recall that $\rho_* = \frac{n\rho}{n+\rho}$. Of course, it is enough to pick ρ and p as small as possible. If $n = 1, 2$, then ρ may be chosen with $\rho_* < 1$ (as a consequence of Corollary 3.6) and we set $p = 1$ in the proof. If $n \geq 3$, then one can always assume that $\rho < p_n = \frac{2n}{n+2}$ and we pick p so that $\rho_* < p < p_n$. Let $f \in C_0^\infty$. We have to establish the weak type estimate

$$(4.14) \qquad |\{x \in \mathbb{R}^n; |L^{1/2}f(x)| > \alpha\}| \leq \frac{C}{\alpha^p} \int |\nabla f|^p,$$

for all $\alpha > 0$. We use the following resolution of $L^{1/2}$:

$$L^{1/2}f = c \int_0^\infty e^{-t^2 L} Lf \, dt$$

where $c = 2\pi^{-1/2}$ is forgotten from now on. It suffices to obtain the result for the truncated integrals $\int_\varepsilon^R \ldots$ with bounds independent of ε, R, and then to let $\varepsilon \downarrow 0$ and $R \uparrow \infty$. For the truncated integrals, all the calculations are justified. We ignore this step and argue directly on $L^{1/2}$. Apply the Calderón-Zygmund decomposition of Lemma 4.12 to f at height α^p and write $f = g + \sum_i b_i$. By construction, $\|\nabla g\|_p \leq c\|\nabla f\|_p$. Interpolating with (4.8) yields $\int |\nabla g|^2 \leq c\alpha^{2-p} \int |\nabla f|^p$. Hence

$$\left|\left\{x \in \mathbb{R}^n; |L^{1/2}g(x)| > \frac{\alpha}{3}\right\}\right| \leq \frac{C}{\alpha^2} \int |L^{1/2}g|^2 \leq \frac{C}{\alpha^2} \int |\nabla g|^2 \leq \frac{C}{\alpha^p} \int |\nabla f|^p$$

where we used the L^2-estimate (2.11) for square roots. To compute $L^{1/2}b_i$, let $r_i = 2^k$ if $2^k \leq \ell_i = \ell(Q_i) < 2^{k+1}$ and set $T_i = \int_0^{r_i} e^{-t^2 L} L \, dt$ and $U_i = \int_{r_i}^\infty e^{-t^2 L} L \, dt$. It is enough to estimate $A = |\{x \in \mathbb{R}^n; |\sum_i T_i b_i(x)| > \alpha/3\}|$ and $B = |\{x \in \mathbb{R}^n; |\sum_i U_i b_i(x)| > \alpha/3\}|$. Let us bound the first term.

First,

$$A \leq |\cup_i 4Q_i| + \left|\left\{x \in \mathbb{R}^n \setminus \cup_i 4Q_i; \left|\sum_i T_i b_i(x)\right| > \frac{\alpha}{3}\right\}\right|,$$

and by (4.10), $|\cup_i 4Q_i| \leq \frac{C}{\alpha^p} \int |\nabla f|^p$.

For the other term, we have

$$\left|\left\{x \in \mathbb{R}^n \setminus \cup_i 4Q_i; \left|\sum_i T_i b_i(x)\right| > \frac{\alpha}{3}\right\}\right| \leq \frac{C}{\alpha^2} \int \left|\sum_i h_i\right|^2$$

with $h_i = \mathbf{1}_{(4Q_i)^c} |T_i b_i|$. To estimate the L^2 norm, we dualize against $u \in L^2$ with $\|u\|_2 = 1$:

$$\int |u| \sum_i h_i = \sum_i \sum_{j=2}^\infty A_{ij}$$

where

$$A_{ij} = \int_{C_j(Q_i)} |T_i b_i| |u|.$$

Let $q = 2$ if $n \leq 2$ and $q = p^* = \frac{np}{n-p}$ if $n \geq 3$. Observe that $\rho < q \leq 2$ and that $p \leq q \leq p^*$.

Since $\rho < q \leq 2$, the family $(tLe^{-tL})_{t>0}$ satisfies $L^q - L^2$ off-diagonal estimates on combining Proposition 3.2 and Proposition 3.15. Hence, using also $r_i \sim \ell_i$,

$$\|e^{-t^2 L} Lb_i\|_{L^2(C_j(Q_i))} \leq \frac{C}{t^{\gamma+2}} e^{-\frac{c 4^j r_i^2}{t^2}} \|b_i\|_q$$

where $\gamma = \gamma_q = |\frac{n}{2} - \frac{n}{q}|$. By Poincaré-Sobolev inequality (since $p \leq q \leq p^*$) and (4.9),
$$\|b_i\|_q \leq c\ell_i^{1-(\frac{n}{p}-\frac{n}{q})} \|\nabla b_i\|_p \leq c\alpha \ell_i^{1+\frac{n}{q}}.$$
Hence, by Minkowski integral inequality, for some appropriate positive constants C, c,
$$\|T_i b_i\|_{L^2(C_j(Q_i))} \leq \int_0^{r_i} \|e^{-t^2 L} L b_i\|_{L^2(C_j(Q_i))}\, dt$$
$$\leq C\alpha e^{-c4^j} \ell_i^{\frac{n}{2}},$$
Now remark that for any $y \in Q_i$ and any $j \geq 2$,
$$\left(\int_{C_j(Q_i)} |u|^2\right)^{1/2} \leq \left(\int_{2^{j+1} Q_i} |u|^2\right)^{1/2} \leq (2^{n(j+1)} |Q_i|)^{1/2} (M(|u|^2)(y))^{1/2}.$$
Applying Hölder inequality, one obtains
$$A_{ij} \leq C\alpha 2^{nj/2} e^{-c4^j} \ell_i^n \left(M(|u|^2)(y)\right)^{1/2}.$$
Averaging over Q_i yields
$$A_{ij} \leq C\alpha 2^{nj/2} e^{-c4^j} \int_{Q_i} \left(M(|u|^2)(y)\right)^{1/2} dy.$$
Summing over $j \geq 2$ and i, we have
$$\int |u| \sum_i h_i \leq C\alpha \int \sum_i \mathbf{1}_{Q_i}(y) \left(M(|u|^2)(y)\right)^{1/2} dy.$$
As in the proof of Theorem 1.1 using finite overlap (4.11) of the cubes Q_i and Kolmogorov's inequality, one obtains
$$\int |u| \sum_i h_i \leq C' N\alpha |\cup_i Q_i|^{1/2} \||u|^2\|_1^{1/2}.$$
Hence
$$\left|\left\{x \in \mathbb{R}^n \setminus \cup_i 4Q_i; \left|\sum_i T_i b_i(x)\right| > \frac{\alpha}{3}\right\}\right| \leq C |\cup_i Q_i| \leq \frac{C}{\alpha^p} \int |\nabla f|^p$$
by (4.11) and (4.10).

It remains to handling the term B. Using functional calculus for L one can compute U_i as $r_i^{-1} \psi(r_i^2 L)$ with ψ the holomorphic function on the sector $|\arg z| < \frac{\pi}{2}$ given by

(4.15) $$\psi(z) = \int_1^\infty e^{-t^2 z} z\, dt.$$

It is easy to show that $|\psi(z)| \leq C|z|^{1/2} e^{-c|z|^2}$, uniformly on subsectors $|\arg z| \leq \mu < \frac{\pi}{2}$.

We invoke the following lemma proved by duality in Chapter 6. [4]

---

[4] It can also be proved directly using the vector-valued extension of Theorem 1.1.

LEMMA 4.14. *If $\rho \in \mathcal{M}_-$ then for $\rho < q \le 2$*

$$(4.16) \quad \left\| \sum_{k \in \mathbb{Z}} \psi(4^k L) \beta_k \right\|_q \lesssim \left\| \left(\sum_{k \in \mathbb{Z}} |\beta_k|^2 \right)^{1/2} \right\|_q,$$

whenever the right hand side is finite.

To apply this lemma, observe that the definitions of r_i and U_i yield

$$\sum_i U_i b_i = \sum_{k \in \mathbb{Z}} \psi(4^k L) \beta_k$$

with

$$\beta_k = \sum_{i, r_i = 2^k} \frac{b_i}{r_i}.$$

Using the bounded overlap property (4.11), one has that

$$\left\| \left(\sum_{k \in \mathbb{Z}} |\beta_k|^2 \right)^{1/2} \right\|_q^q \le C \int \sum_i \frac{|b_i|^q}{r_i^q}.$$

By (A.2), and $p \le q \le p^*$, together with $\ell_i \sim r_i$,

$$\int \sum_i \frac{|b_i|^q}{r_i^q} \le C \alpha^q \sum_i |Q_i|.$$

Hence, by (4.10)

$$\left| \left\{ x \in \mathbb{R}^n; \left| \sum_i U_i b_i(x) \right| > \frac{\alpha}{3} \right\} \right| \le C \sum_i |Q_i| \le \frac{C}{\alpha^p} \int |\nabla f|^p.$$

$\square$

4.3. Invertibility

We finish the study of square roots with their invertibility properties on L^p spaces. This is summed up in the following theorem. Recall that $\mathcal{I}(L)$ is the maximal interval of exponents $p \in (1, \infty)$ for which $\nabla L^{-1/2}$ is bounded on L^p.

THEOREM 4.15. *Let $1 < p < \infty$. Then $p \in \mathcal{I}(L)$ if and only if $L^{1/2}$, a priori defined from C_0^∞ into L^2, extends to an isomorphism from $\dot{W}^{1,p}$ onto L^p with $\|L^{1/2} f\|_p \sim \|\nabla f\|_p$. Furthermore, $\mathcal{I}(L)$ is an open interval.*

PROOF. For $p = 2$, the equivalence is a consequence of the solution of the Kato's conjecture (2.11) as recalled in Section 2.4.

For other values of p, one implication is straightforward. Conversely, we know from Theorem 4.9 that $\mathcal{R}(L)$, the maximal interval of exponents $p \in (1, \infty)$ for which one has the *a priori* inequality $\|L^{1/2} f\|_p \lesssim \|\nabla f\|_p$, contains a neighborhood of the closure (in $\mathbb{R}$) of $\mathcal{I}(L)$.

In particular, if $\nabla L^{-1/2}$ is bounded on L^p then for the same p, $L^{1/2}$ can be extended boundedly from $\dot{W}^{1,p}$ into L^p. The isomorphism property is now easy. Indeed, from $\|L^{1/2} f\|_p \sim \|\nabla f\|_p$, we deduce that this extension is one-one and has closed range in L^p. It remains to establish the density of this range. If $g \in L^p \cap L^2$, we have $g = L^{1/2} L^{-1/2} g$ and by $\|\nabla L^{-1/2} g\|_p \le c \|g\|_p < \infty$, we conclude that g is in $L^{1/2}(\dot{W}^{1,p})$. Thus $L^{1/2}(\dot{W}^{1,p})$ contains a dense subspace of L^p.

The openness of $\mathcal{I}(L)$ is a consequence of the following result.

LEMMA 4.16. [5] *Let X^s, Y^s, be two scales of complex interpolation Banach spaces, s describing an open interval I. If $T: X^s \to Y^s$ is bounded for each $s \in I$, then the set of $s \in I$ for which T is an isomorphism from X^s onto Y^s is open.*

Apply this result (changing s to p) with $I = \operatorname{Int} \mathcal{R}(L)$, $X^p = \dot{W}^{1,p}$, $Y^p = L^p$ and $T = L^{1/2}$: we just proved that the set of $p \in \operatorname{Int} \mathcal{R}(L)$ for which $L^{1/2}$ is an isomorphism from $\dot{W}^{1,p}$ onto L^p is $\mathcal{I}(L)$. Thus, $\mathcal{I}(L)$ is open. $\square$

4.4. Applications

We have developed the necessary theory to reobtain all the results mentioned in the Introduction. Let us start with results holding in all generality.

PROPOSITION 4.17. [6] *If $n = 1$, we have $\|L^{1/2}f\|_p \sim \|\frac{d}{dx}f\|_p$ and $L^{1/2}$ extends to an isomorphism from $\dot{W}^{1,p}$ onto L^p when $1 < p < \infty$.*

PROOF. We know from Corollary 3.6 that $p_-(L) = 1$ and from Proposition 3.12 that $q_+(L) = \infty$. Hence $\mathcal{I}(L) = (1, \infty)$ and the conclusion follows from Theorem 4.15. $\square$

PROPOSITION 4.18. [7] *If $n = 2$, we have $\|\nabla L^{-1/2}f\|_p \lesssim \|f\|_p$ for $1 < p < 2 + \varepsilon$ and $\|L^{1/2}f\|_p \lesssim \|\nabla f\|_p$ for $1 < p < \infty$. Furthermore, $L^{1/2}$ extends to an isomorphism from $\dot{W}^{1,p}$ onto L^p when $1 < p < 2 + \varepsilon$.*

PROOF. We know from Corollary 3.6 that $p_-(L) = 1$ and from Corollary 3.10 that $q_+(L) > 2$. Hence, $\mathcal{I}(L) = (1, 2 + \varepsilon)$ and we have the isomorphism property from Theorem 4.15.

Now, $r_-(L) = 1$ and $r_+(L) \geq p_+(L) = (p_-(L^*))' = \infty$. Hence $\|L^{1/2}f\|_p \lesssim \|\nabla f\|_p$ for $1 < p < \infty$ by Theorem 4.9. $\square$

PROPOSITION 4.19. [8] *If $n \geq 3$, then $\|\nabla L^{-1/2}f\|_p \lesssim \|f\|_p$ for $\frac{2n}{n+2} - \varepsilon < p < 2 + \varepsilon'$ and $\|L^{1/2}f\|_p \lesssim \|\nabla f\|_p$ for $\sup(1, \frac{2n}{n+4} - \varepsilon_1) < p < \frac{2n}{n-2} + \varepsilon_1'$ with $\varepsilon, \varepsilon', \varepsilon_1, \varepsilon_1' > 0$ depending on dimension and the ellipticity constants of L. Furthermore, $L^{1/2}$ extends to an isomorphism from $\dot{W}^{1,p}$ onto L^p when $\frac{2n}{n+2} - \varepsilon < p < 2 + \varepsilon'$.*

PROOF. We know from Corollary 3.6 that $p_-(L) < \frac{2n}{n+2}$ and from Corollary 3.10 that $q_+(L) > 2$. Hence, $\mathcal{I}(L) = (\frac{2n}{n+2} - \varepsilon, 2 + \varepsilon')$ and the isomorphism property follows from Theorem 4.15. By Theorem 4.9, $r_-(L) \leq \sup(1, (p_-(L))_*)$ and $r_+(L) \geq p_+(L) = (p_-(L^*))'$. Thus $\|L^{1/2}f\|_p \lesssim \|\nabla f\|_p$ for $\sup(1, \frac{2n}{n+4} - \varepsilon_1) < p < \frac{2n}{n-2} + \varepsilon_1'$. $\square$

REMARK. If $n \geq 2$, the bound $p < 2+\varepsilon'$ for the Riesz transform L^p boundedness is sharp as $q_+(L) > 2$ is optimal (see Section 3.5).

---

[5] This is due to I. Sneiberg ([60]).

[6] This is first proved in [18], Theorem A. More is proved there: in that $a\frac{d}{dx}L^{-1/2}$ is a Calderón-Zygmund operator and the boundedness properties at $p = 1$ and $p = \infty$ are studied.

[7] This follows from [17], Chapter IV, Theorem 1, using [15], Theorem 3.5 and [10], Theorem 1.4.

[8] The first inequality is proved in [21], [42] for the range $\frac{2n}{n+2} < p < 2$ and the extension to this larger range is in [6], the second inequality is in [42] for the range $2 < p < \frac{2n}{n-2}$, and in [6] for the rest of the range.

A new fact is the following negative result.

COROLLARY 4.20. *If $n \geq 5$, there exists $p \in (1,2)$ and an operator L for which $\nabla L^{-1/2}$ is not bounded on L^p.*

PROOF. If $n \geq 5$, there exists an operator L for which $p_-(L) > 1$ (see Section 3.5). It remains to invoke Theorem 4.2. □

Let us come to results where further hypotheses may be taken on L.

COROLLARY 4.21. [9] *If $n \geq 3$ and L has **real** coefficients, then $\|\nabla L^{-1/2} f\|_p \lesssim \|f\|_p$ for $1 < p < 2 + \varepsilon$ and $\|L^{1/2} f\|_p \lesssim \|\nabla f\|_p$ for $1 < p < \infty$. Furthermore, $L^{1/2}$ extends to an isomorphism from $\dot{W}^{1,p}$ onto L^p when $1 < p < 2 + \varepsilon$.*

PROOF. The semigroup $(e^{-tL})_{t>0}$ is contracting on L^1. By Proposition 3.2, part 1, this implies L^1-L^2 boundedness. By Proposition 4.3, we obtain $\|\nabla L^{-1/2} f\|_p \lesssim \|f\|_p$ for $1 < p < 2$. The rest of the proof is as the one of Proposition 4.18. □

REMARK. If $n \geq 2$ and L has continuous and periodic coefficients with common period then a careful spectral analysis of the semigroup yields that the Riesz transform is bounded on L^p for all $p \in (1, \infty)$. [10] As a consequence, one obtains uniform gradient bounds on the heat operator from L^p to L^p, for any $1 < p < \infty$.

REMARK. If $n \geq 2$ and L has real, Hölder continuous and quasiperiodic coefficients satisfying Koslov condition then by homogenization techniques, the Riesz transform is proved to be bounded on L^p for all $p \in (1, \infty)$. [11] As a consequence, one obtains uniform gradient bounds on the heat operator from L^p to L^p, for any $1 < p < \infty$.

REMARK. If $n \geq 2$ and L has almost-periodic coefficients then the semigroup is $L^1 - L^2$ bounded (in fact, much more is true), which is enough to conclude that the Riesz transform is bounded on L^p for all $p \in (1, 2)$. The situation for $p > 2$ is unclear. [12]

4.5. Riesz transforms and Hodge decomposition

An L^p Hodge decomposition adapted to the operator L consists in writing a field $f \in L^p$ into the sum $g + \nabla h$ where the fields $g, \nabla h \in L^p$ with $\|g\|_p + \|\nabla h\|_p \leq c\|f\|_p$ and $\operatorname{div}(Ag) = 0$. This amounts to the boundedness of the Hodge projection $\nabla L^{-1} \operatorname{div}$ on L^p, that is an inequality of the type

(4.17) $$\|\nabla L^{-1} \operatorname{div} f\|_p \lesssim \|f\|_p.$$

Indeed, formally, one has
$$\nabla L^{-1} \operatorname{div}(Af) = \nabla h.$$

Alternately, it can be seen as the boundedness for the second order Riesz transform, $\nabla L^{-1/2} (\nabla (L^*)^{-1/2})^*$.

For $p = 2$, the inequality (4.17) in automatic by construction of L. For $p \neq 2$, we see here the connection to Riesz transforms estimates. Before, we put aside the one dimensional case.

---

[9] This is first proved in [**17**], Chapter 4, using [**10**], Theorem 1.4 and Aronson's estimates in [**5**].
[10] This is in [**35**], Theorem 1.1.
[11] This is in [**1**], Theorem 1.2.
[12] This is in [**32**], Theorem 1.1.

PROPOSITION 4.22. *If $n = 1$, $\frac{d}{dx} L^{-1} \frac{d}{dx}$ extends to a bounded operator on L^p, $1 < p < \infty$, equal to the negative of the operator of pointwise multiplication with $\frac{1}{a(x)}$.*

PROOF. Let D be the space of compactly supported and C^1 functions f with $\int_{-\infty}^{\infty} \frac{f(t)}{a(t)} dt = 0$. It is easy to see that D is dense in L^p when $1 < p < \infty$. For f in D, the unique solution in $\dot{W}^{1,2}$ of $Lu = \frac{df}{dx}$ is given by

$$u(x) = \int_x^{\infty} \frac{f(t)}{a(t)} dt, \quad x \in \mathbb{R}.$$

We have, $u'(x) = -\frac{f(x)}{a(x)}$ and $u' \in L^p$ with $\|u'\|_p \leq \|\frac{1}{a}\|_\infty \|f\|_p$ since $\frac{1}{a}$ is a bounded function. Thus the boundedness property holds from D into L^p and the conclusion of the proposition follows readily. □

Recall that Riesz transforms inequalities hold in the full range $1 < p < \infty$ when $n = 1$.

We now restrict our attention to dimensions $n \geq 2$. In that case, (4.17) is always valid for $|\frac{1}{2} - \frac{1}{p}| < \varepsilon$, the value of ε depending on the ellipticity constant of L and dimension.

THEOREM 4.23. (1) *Let $2 < p < \infty$ and $n \geq 2$. Then ∇L^{-1} div is bounded on L^p if and only if $\nabla L^{-1/2}$ is bounded on L^p.*
(2) *Let $1 < p < 2$ and $n \geq 3$. If ∇L^{-1} div is bounded on L^p then $\nabla L^{-1/2}$ is bounded on L^q for $\sup(1, p_*) < q < 2$.*

PROOF. Let us begin the argument with the second statement. Let $1 < p < 2$ and $n \geq 3$. Assume $\|\nabla L^{-1} \operatorname{div} f\|_p \lesssim \|f\|_p$ for all $f \in L^p$. Using interpolation with the corresponding L^2-inequality, there is no loss of generality to assume $p_* > 1$. By Sobolev embeddings we deduce that $L^{-1} : L^{p_*} \to L^{p^*}$ for any such p. Take p_0 with $p_* < p_0 < p$. Define p_k by $p_k = ((p_{k-1})_*)^*$ if $p_0 \leq 2_*$ and stop when $2_* < p_k \leq 2^*$.

If $k = 0$, then we already know that $(e^{-tL})_{t>0}$ is $L^{p_0} - L^2$ bounded, hence $p_0 \in \mathcal{J}_-(L)$. By Theorem 4.2, we conclude that $(p_0, 2) \subset \mathcal{I}_-(L)$. Since p_0 was arbitrary, we have shown that $(p_*, 2) \subset \mathcal{I}_-(L)$ in this case.

Assume now, that $k \geq 1$. By construction, L^{-k} is bounded from L^{p_0} to L^{p_k}. Since $2_* \leq p_k \leq 2^*$, $(e^{-tL})_{t>0}$ is an analytic semigroup on L^{p_k} by Proposition 3.15. As in Step 2 of the proof of Theorem 4.2, we obtain that e^{-tL} is bounded from L^{p_0} to L^{p_k}. We also have that $(e^{-tL})_{t>0}$ satisfies L^{p_k} off-diagonal estimates by interpolation between L^2 off-diagonal estimates and L^r boundedness for r chosen so that p_k is between 2 and r. By interpolation again, we deduce that if $p_0 < q < p_k$, $(e^{-tL})_{t>0}$ satisfies $L^q - L^{p_k}$ off-diagonal estimates. Using Lemma 3.3, this implies that $(e^{-tL})_{t>0}$ is L^q bounded. Hence, $q \in \mathcal{J}_-(L)$. We conclude as above that $(p_*, 2) \subset \mathcal{I}_-(L)$.

Let us now consider the case $p > 2$. Assume $\|\nabla L^{-1} \operatorname{div} f\|_p \lesssim \|f\|_p$. We first claim that $\|(L^*)^{1/2} f\|_{p'} \lesssim \|\nabla f\|_{p'}$. Indeed, this always holds if $n = 2$ by Proposition 4.18. If $n \geq 3$, by duality we have $\|\nabla (L^*)^{-1} \operatorname{div} f\|_{p'} \lesssim \|f\|_{p'}$ and the preceeding case tells us that $\nabla (L^*)^{-1/2}$ is $L^{p'}$ bounded. Applying Theorem 4.9 proves the claim.

Secondly, let $h \in L^{p'} \cap \dot{W}^{1,2}$. Since $\operatorname{div} h \in L^2$, we have that

$$(L^*)^{-1/2} \operatorname{div} h = (L^*)^{1/2} (L^*)^{-1} \operatorname{div} h,$$

hence
$$\|(L^*)^{-1/2}\operatorname{div} h\|_{p'} = \|(L^*)^{1/2} L^{*-1}\operatorname{div} h\|_{p'} \lesssim \|\nabla (L^*)^{-1}\operatorname{div} h\|_{p'} \lesssim \|h\|_{p'}.$$

Since $L^{p'} \cap \dot{W}^{1,2}$ is dense in $L^{p'}$, we obtain the $L^{p'}$-boundedness of $(L^*)^{-1/2}\operatorname{div}$ which, by duality, means the L^p-boundedness of the Riesz transform $\nabla L^{-1/2}$.

For the converse, we deduce from the characterization of $\mathcal{I}(L)$ and the relation between $p_\pm(L)$ and $q_\pm(L)$ that the L^p-boundedness of the Riesz transform $\nabla L^{-1/2}$ and $p > 2$ imply the $L^{p'}$-boundedness of the Riesz transform $\nabla (L^*)^{-1/2}$, which by duality means that $L^{-1/2}\operatorname{div}$ is bounded on L^p. Hence, $\nabla L^{-1}\operatorname{div} = \nabla L^{-1/2} L^{-1/2} \operatorname{div}$ is bounded on L^p. □

REMARK. By Proposition 4.18, the second statement is meaningless if $n = 2$, hence the assumption $n \geq 3$. Its converse is false for any $p < 2$. For example, for any $\varepsilon > 0$, there exist real symmetric operators L such that $\|\nabla L^{-1}\operatorname{div} f\|_p \lesssim \|f\|_p$ only when $|\frac{1}{2} - \frac{1}{p}| < \varepsilon$, yet its associated Riesz transform is L^p bounded when $1 < p < 2$.

REMARK. Let us describe a geometric interpretation of the reverse inequalities for square roots. Let $\mathcal{H}_p(L) = \{g \in L^p; \operatorname{div}(Ag) = 0\}$ and $\mathcal{G}_p = \nabla(\dot{W}^{1,p})$. These are closed subspaces of L^p. The Hodge decomposition in L^p is equivalent to having $\mathcal{H}_p(L) + \mathcal{G}_p = L^p$ as a topological direct sum. By duality, it is also equivalent to $\mathcal{H}_{p'}(L^*) + \mathcal{G}_{p'} = L^{p'}$ is a topological direct sum.

Assume that $\|L^{1/2} f\|_p \lesssim \|\nabla f\|_p$ holds. Then, by duality $(L^*)^{1/2}$ is bounded from $L^{p'}$ into $\dot{W}^{-1,p'}$. If $\nabla h \in \mathcal{G}_{p'}$ then $(L^*)^{-1/2}\operatorname{div}(A^*\nabla h) = (L^*)^{1/2} h$ makes sense. Hence, the restriction of $(L^*)^{-1/2}\operatorname{div}(A^*\cdot)$ to $\mathcal{G}_{p'}$ is bounded into $L^{p'}$. If $g \in \mathcal{H}_{p'}(L^*)$ then $(L^*)^{-1/2}\operatorname{div}(A^*g) = 0$, hence the restriction of $(L^*)^{-1/2}\operatorname{div}(A^*\cdot)$ to $\mathcal{H}_{p'}(L^*)$ is bounded into $L^{p'}$ (without any hypothesis). Conversely, these two facts imply $\|L^{1/2} f\|_p \lesssim \|\nabla f\|_p$. Thus, this inequality means that $(L^*)^{-1/2}\operatorname{div}(A^*\cdot)$ is bounded on closed subspaces of $L^{p'}$ into $L^{p'}$ even without knowing whether they are in direct sum in $L^{p'}$.

If the topological direct sum holds then $(L^*)^{-1/2}\operatorname{div}(A^*\cdot)$ is bounded on $L^{p'}$, hence $\nabla L^{-1/2}$ is bounded on L^p. This is what we proved above.

This also illustrates why $\|L^{1/2} f\|_p \lesssim \|\nabla f\|_p$ is possible even when $\nabla L^{-1/2}$ is not bounded on L^p.

Let us finish with another identification of $q_+(L)$.

COROLLARY 4.24. *$q_+(L)$ is the supremum of exponents p for which one has the Hodge decomposition in L^p, or alternatively for which L extends to an isomorphism from $\dot{W}^{1,p}$ onto $\dot{W}^{-1,p}$.*[13] *The interval of values of p for which this holds is, therefore, $((q_+(L^*))', q_+(L))$.*

PROOF. If $n = 1$ we have $q_+(L) = \infty$ and the Hodge projections are bounded on all L^p spaces as recalled above. If $n \geq 2$, this follows right away from the previous theorem and the fact that $q_+(L)$ is the supremum of $\mathcal{I}(L)$. □

---

[13] See Lemma 3.4.

CHAPTER 5

Riesz transforms and functional calculi

In this chapter, we present the theorem of Blunck & Kunstmann concerning H^∞ functional calculus on L^p spaces. We also discuss Hardy-Littlewood-Sobolev inequalities. Combining this with Riesz transform estimates, we obtain a family of inequalities which we summed up in what we call the Hardy-Littlewood-Sobolev-Kato diagram.

5.1. Blunck & Kunstmann's theorem

Let L be as usual and ω be the type of L defined in (2.4). We know that L admits a bounded holomorphic functional calculus on L^2.

Let $p \in (1, \infty)$. We say that L has a bounded holomorphic functional calculus on L^p if one has the following property: for any $\mu \in (\omega, \pi)$ and any φ holomorphic and bounded in Σ_μ, and all $f \in L^2 \cap L^p$

$$\|\varphi(L)f\|_p \leq c\|\varphi\|_\infty \|f\|_p$$

the constant c depending only on p, ω and μ. The operator $\varphi(L)$ is defined on L^p by density. [1]

We define $\mathcal{H}(L)$ as the sets of those exponents p with the above property. By interpolation, these sets are intervals (if nonempty).

Recall that $\mathcal{J}(L)$ is the set of exponents p for which $(e^{-tL})_{t>0}$ is L^p bounded and $\mathcal{J}(L) \cap [1,2) = \mathcal{J}_-(L)$.

THEOREM 5.1. *The sets $\mathcal{J}(L)$ and $\mathcal{H}(L)$ have same interiors.*

By Theorem 4.2, this gives us

COROLLARY 5.2. *The sets $\mathcal{I}_-(L)$ and $\mathcal{H}_-(L)$ have same interiors.*

PROOF OF THEOREM 5.1. It is enough to show that $\mathcal{J}_-(L)$ and $\mathcal{H}_-(L)$ have same interiors as one can use duality in this context.

Let $p < 2$. If L has a bounded holomorphic functional calculus on L^p then the semigroup $(e^{-tL})_{t>0}$ is L^p bounded. Hence $\mathcal{H}_-(L) \subset \mathcal{J}_-(L)$.

Conversely, it is enough to show that $p_0 \in \text{Int}\,\mathcal{K}_-(L)$, that is $(e^{-zL})_{z \in \Sigma_\beta}$ satisfies $L^{p_0} - L^2$ off-diagonal estimates for any $\beta \in (0, \frac{\pi}{2} - \omega)$ (see Proposition 3.15), implies that $\varphi(L)$ is weak-type (p_0, p_0) whenever φ is holomorphic and bounded in Σ_μ for $\mu > \omega$ and we may choose $\mu < \frac{\pi}{2}$ by convenience. To this end, we apply Theorem 1.1 to $T = \varphi(L)$. It is enough to assume further on φ the technical condition (2.5). [2]

We set $A_r = I - (I - e^{-r^2 L})^m$, $r > 0$, for some large integral number m.

---

[1] Strictly speaking, one should incorporate a statement about convergence of operators to allow limiting procedures. In fact, this follows from the L^2-functional calculus and density.

[2] This follows from the Convergence Lemma in [53].

Assume $p_0 < 2$ is such that $(e^{-zL})_{z \in \Sigma_\beta}$ satisfies $L^{p_0} - L^2$ off-diagonal estimates for any $\beta \in (0, \frac{\pi}{2} - \omega)$. It is enough to check (1.1) as (1.2) is granted from the assumption on the semigroup. Our goal is to establish the inequality

$$(5.1) \quad \left(\frac{1}{|2^{j+1}B|} \int_{C_j(B)} |\varphi(L)(I - e^{r^2 L})^m f|^2\right)^{1/2} \leq g(j) \left(\frac{1}{|B|} \int_B |f|^{p_0}\right)^{1/p_0}$$

for all ball B with r the radius of B and all f supported in B and all $j \geq 2$, with $\sum 2^{nj} g(j) < \infty$. As before, $C_j(B)$ denotes the ring $2^{j+1}B \setminus 2^j B$.

To do this, let $\psi(z) = \varphi(z)(1 - e^{-r^2 z})^m$ so that $\psi(L) = \varphi(L)(I - e^{r^2 L})^m$. Represent $\psi(L)$ using the representation formula (2.6). Using the exact form of ψ and the definition of $\eta_\pm$, it is easy to obtain

$$|\eta_\pm(z)| \leq C \|\varphi\|_\infty |z|^{-1} \inf\left(1, r^{2m}|z|^{-m}\right), \quad z \in \Gamma_\pm.$$

Hence, by Minkowski integral inequality in (2.6), the hypothesis on p_0 and on the support of f, $\|\varphi(L)(I - e^{r^2 L})^m f\|_{L^2(C_j(B))}$ is bounded above by

$$C \int_{\Gamma_+} e^{-\frac{c 4^j r^2}{|z|}} \frac{1}{|z|^{\gamma_{p_0}/2}} \inf\left(1, r^{2m}|z|^{-m}\right) \frac{d|z|}{|z|} \|f\|_{p_0}$$

plus the term corresponding to integration on Γ_-. A calculation gives us a bound

$$C r^{-\gamma_{p_0}} 2^{-j(\gamma_{p_0} + 2m)} \|f\|_{p_0}.$$

Using the value of γ_{p_0} we obtain (5.1) with $g(j) = C 2^{-nj/2} 2^{-j(\gamma_{p_0} + 2m)}$. Choosing m with $\gamma_{p_0} + 2m > n/2$ concludes the argument. $\square$

5.2. Hardy-Littlewood-Sobolev estimates

We have the following result.

PROPOSITION 5.3. *Let $p_-(L) < p < q < p_+(L)$. Then $L^{-\alpha}$ is bounded from L^p into L^q provided*

$$\alpha = \frac{1}{2}\left(\frac{n}{p} - \frac{n}{q}\right).$$

PROOF. We first observe that with the following choice of p and q then the family $(e^{-tL})_{t>0}$ is $L^p - L^q$ bounded. Indeed, we know it if $p = 2$ or if $q = 2$ by Proposition 3.2. If $p < 2 < q$, it suffices to use composition. If $p < q < 2$, interpolate between L^p boundedness and $L^p - L^2$ boundedness. And if $2 < p < q$, interpolate between L^q boundedness and $L^2 - L^q$ boundedness.

Next, by L^2 functional calculus, we have

$$f = \frac{1}{\Gamma(\alpha)} \int_0^\infty t^{\alpha - 1} L^\alpha e^{-tL} f \, dt$$

for all $f \in L^2$, where the integrals $\int_\varepsilon^R \ldots$ converge in L^2 as $\varepsilon \downarrow 0$ and $R \uparrow \infty$. Set $T_{\varepsilon,R} = \Gamma(\alpha)^{-1} \int_\varepsilon^R t^{\alpha-1} e^{-tL} \, dt$. Let $f \in L^2 \cap L^p$ with $\|f\|_p = 1$. Fix $\alpha = \frac{1}{2}(\frac{n}{p} - \frac{n}{q})$. For $a > 0$ and $p < q_0 < q < q_1 < q_+(L)$, we easily obtain with uniform constant C,

$$\left\| \int_\varepsilon^a t^{\alpha-1} e^{-tL} f \, dt \right\|_{q_1} \leq C a^{\frac{1}{2}(\frac{n}{q} - \frac{n}{q_1})}$$

5.2. HARDY-LITTLEWOOD-SOBOLEV ESTIMATES

and
$$\left\| \int_a^R t^{\alpha-1} e^{-tL} f \, dt \right\|_{q_0} \leq C a^{\frac{1}{2}(\frac{n}{q_0} - \frac{n}{q})}.$$

Hence, by the argument of Marcinkiewic interpolation theorem, we have if $\lambda > 0$
$$|\{x \in \mathbb{R}^n; \ |T_{\varepsilon,R} f(x)| > \lambda\}| \leq C \lambda^{-q_1} a^{\frac{q_1}{2}(\frac{n}{q} - \frac{n}{q_1})} + C \lambda^{-q_0} a^{\frac{q_0}{2}(\frac{n}{q_0} - \frac{n}{q})}.$$

Choosing $a^{\frac{n}{2q}} = \lambda$ yields,
$$|\{x \in \mathbb{R}^n; \ |T_{\varepsilon,R} f(x)| > \lambda\}| \leq 2 C \lambda^{-q}.$$

Hence, $T_{\varepsilon,R} f$ belongs to the Lorentz space $L^{q,\infty}$. Since this holds for all q as above, by interpolation again, we conclude that $T_{\varepsilon,R} f \in L^q$ and
$$\|T_{\varepsilon,R} f\|_q \leq C \|f\|_p$$

whenever $f \in L^2 \cap L^p$ with uniform constant with respect to ε, R.

It remains to pass to the limit. If $0 \leq \alpha \leq 1/2$, we claim that L^α defines an isomorphism between the fractional Sobolev space $\dot{H}^{2\alpha}$ (defined as the closure of C_0^∞ for the seminorm $\|f\|_{\dot{H}^{2\alpha}} \equiv \|(-\Delta)^\alpha f\|_2$) and L^2 with
$$\|L^\alpha f\|_2 \sim \|f\|_{\dot{H}^{2\alpha}}.$$

Indeed, if $\Re \alpha = 0$, $z \mapsto z^\alpha$ is bounded holomorphic in Σ_π, hence
$$\|L^\alpha f\|_2 \sim \|f\|_2$$

with implicit constants growing at most like $C e^{c|\Im \alpha|}$ with some $0 < c < \pi/2$. For $\Re \alpha = 1/2$, combining the square root problem (2.11) with the L^2 functional calculus, we have
$$\|L^\alpha f\|_2 \sim \|\nabla f\|_2$$

with the same growth as above for the implicit constants. The claim follows by complex interpolation. By construction, if $f \in L^2$ then
$$\lim_{\varepsilon \downarrow 0, R \uparrow \infty} \|T_{\varepsilon,R} f - L^{-\alpha} f\|_{\dot{H}^{2\alpha}},$$

hence, for α sufficiently small so that Sobolev embeddings $\dot{H}^{2\alpha} \subset L^r$ applies with $r < \infty$, we also have the convergence in L^r.

Now, if $f \in L^2 \cap L^p$, we combine the uniform bound of $T_{\varepsilon,R} f$ in L^q and its convergence in L^r to conclude for the $L^p - L^q$ bounded extension of $L^{-\alpha}$ for all small positive α.

We obtain all possible values of α by writing $L^{-\alpha} = (L^{-\alpha/k})^k$ for k large enough. □

REMARK. The isomorphism property used in the proof holds $0 \leq \alpha < 1/2$ without knowing the solution of the Kato square root problem using that the domain of L^α is an interpolation space between the form domain (*i.e.* $W^{1,2}$) and L^2. [3]

---

[3] This complex interpolation result follows from Kato [45] and Lions [49].

5.3. The Hardy-Littlewood-Sobolev-Kato diagram

In the plane $\{(\frac{1}{p}, s)\}$, where p is a Lebesgue exponent and s a regularity index, we introduce a convex set on which we have a rule for boundedness from $\dot{W}^{s,p}$ to $\dot{W}^{\sigma,q}$ of functions of L. We call it the Hardy-Littlewood-Sobolev-Kato diagram of L because it includes all previously seen estimates from functional calculus and Kato type inequalities (see below).

Pick exponents p_0, q_0, p_1, q_1 as follows:
$$q_-(L) = p_-(L) < p_0 < 2 < q_0 < q_+(L)$$
and
$$q_-(L^*) = p_-(L^*) < p_1 < 2 < q_1 < q_+(L^*).$$
Hence, p_0, q_0 (resp. p_1, q_1) are in the range of L^p-boundedness of the Riesz transform $\nabla L^{-1/2}$ (resp. $\nabla (L^*)^{-1/2}$). Consider the closed convex polygon $\mathcal{P} = ABDFEC$ in the plane $\{(\frac{1}{p}, s)\}$ with $A = (\frac{1}{q_0}, 1)$, $B = (\frac{1}{p_0}, 1)$, $D = (\frac{1}{p_0}, 0)$ $F = (\frac{1}{(q_1)'}, -1)$, $E = (\frac{1}{(p_1)'}, -1)$ and $C = (\frac{1}{(p_1)'}, 0)$. See Figure 1. Let $M = (\frac{1}{p}, s), N = (\frac{1}{q}, \sigma)$ in $\mathcal{P}$. We call $\overrightarrow{MN}$ an authorized arrow if $p \leq q$. Set in this case,
$$\alpha(M, N) = \frac{\sigma - s}{2} + \frac{1}{2}\left(\frac{n}{p} - \frac{n}{q}\right).$$

In other words, we are allowed to move in $\mathcal{P}$ horizontally from the right to the left, vertically up and down and all possible combinations. In fact, an accurate correspondance would be the convex three dimensional set of authorized arrows $\overrightarrow{MN}$ between two parrallel copies of $\mathcal{P}$ in $\mathbb{R}^3$ whenever M in the first copy and N in the second copy.

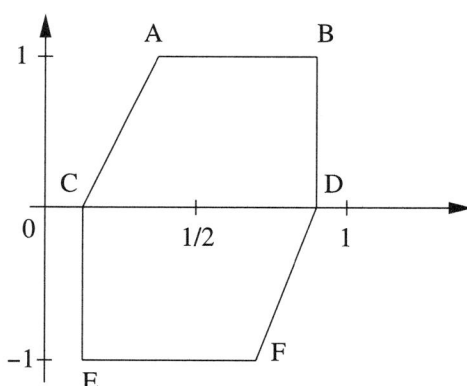

FIGURE 1. **The Hardy-Littlewood-Sobolev-Kato diagram $\mathcal{P}$.**

THEOREM 5.4. *Let $\mu \in (\omega, \pi)$. For any $M = (\frac{1}{p}, s), N = (\frac{1}{q}, \sigma) \in \mathcal{P}$ with $p \leq q$ and any $\varphi \in \mathcal{F}(\Sigma_\mu)$,*
$$\|\varphi(L)f\|_{\dot{W}^{\sigma,q}} \lesssim \|z^{\alpha(M,N)}\varphi\|_{L^\infty(\Sigma_\mu)}\|f\|_{\dot{W}^{s,p}},$$
provided the quantity $\|z^{\alpha(M,N)}\varphi\|_{L^\infty(\Sigma_\mu)}$ is finite. The implicit constant depends on $\mu, \omega, p, q, s, \sigma$.

A few explanations are necessary. First, for $0 < s$, $\dot{W}^{s,p}$ is defined as the closure of C_0^∞ for $\|(-\Delta)^{s/2}f\|_p$ and $\dot{W}^{-s,p'}$ is its dual space. If $0 < \alpha < 1$, $z \mapsto z^\alpha$ is the analytic continuation in Σ_π of $t \mapsto t^\alpha$ defined on $(0, \infty)$. If $k < \alpha < k+1$ with k integral number, then $z \mapsto z^\alpha = z^k z^{\alpha - k}$ is analytic in Σ_π. Hence, if $\varphi \in \mathcal{F}(\Sigma_\mu)$ then $z^\alpha \varphi \in \mathcal{F}(\Sigma_\mu)$ and $L^\alpha \varphi(L)$ is well-defined (See Section 2.2). Also, we shall write *a priori* inequalities for suitable functions f and we shall leave to the reader the care of providing the density arguments.

When $s = \sigma = 0$, this includes $L^p - L^q$ estimates for negative fractional powers of L (Hardy-Littlewood-Sobolev estimates). For $p = q$, we have a comparison between fractional powers of L with the ones of the Laplacian $-\Delta$ (Kato type estimates): the operators $(-\Delta)^{\sigma/2} L^{(s-\sigma)/2} (-\Delta)^{-s/2}$ are bounded on L^p provided $(\frac{1}{p}, s)$ and $(\frac{1}{p}, \sigma)$ belong to $\mathcal{P}$. More precisely

COROLLARY 5.5. *If $(\frac{1}{p}, s)$ and $(\frac{1}{p}, \sigma)$ belong to $\mathcal{P}$, then $L^{(s-\sigma)/2}$ extends to an isomorphism from $\dot{W}^{s,p}$ onto $\dot{W}^{\sigma,p}$*

PROOF. The previous remark means that

$$\|L^{(s-\sigma)/2} f\|_{\dot{W}^{\sigma,p}} \sim \|f\|_{\dot{W}^{s,p}}$$

which implies that the extension from C_0^∞ to $\dot{W}^{s,p}$ is bounded and one-to-one. Since the same thing is true for L^* in the dual range, we conclude that this extension is onto by classical arguments. □

In particular we recover the simultaneous L^p boundedness of the Riesz transform and the Hodge projection when $p > 2$. We pursue the discussion on Kato type estimates in the next section.

PROOF OF THEOREM 5.4. By convexity and complex interpolation, it suffices to prove the result for the following extremal authorized arrows:

(1) the minimal ones: $\overrightarrow{MM}$ where M is one of the six vertices of $\mathcal{P}$.
(2) the maximal horizontal ones: $\overrightarrow{BA}$, $\overrightarrow{DC}$, $\overrightarrow{FE}$.
(3) the maximal vertical ones: $\overrightarrow{DB}$, $\overrightarrow{FH}$, $\overrightarrow{GA}$, $\overrightarrow{EC}$ and their respective opposite where $G = (\frac{1}{p_0}, -1)$ and $H = (\frac{1}{(q_1)'}, 1)$.

Step 1. Minimal arrows.

- $\overrightarrow{AA}$: Write

$$\|\nabla \varphi(L) f\|_{q_0} \lesssim \|L^{1/2} \varphi(L) f\|_{q_0} \sim \|\varphi(L) L^{1/2} f\|_{q_0} \lesssim \|\varphi\|_\infty \|L^{1/2} f\|_{q_0} \lesssim \|\nabla f\|_{q_0}.$$

The first inequality holds because the Riesz transform is L^{q_0} bounded, the second by the commutative property of the functional calculus, the third by Theorem 5.1 and the last by the reverse inequalities at q_0.

- $\overrightarrow{BB}$: Same as $\overrightarrow{AA}$ by changing q_0 to p_0.
- $\overrightarrow{CC}$: We have $(p_1)' \in \text{Int}\,\mathcal{J}(L)$, hence Theorem 5.1 applies and yields the inequality $\|\varphi(L) f\|_{(p_1)'} \lesssim \|\nabla f\|_{(p_1)'}$.
- $\overrightarrow{DD}$: Same as $\overrightarrow{CC}$ by changing $(p_1)'$ to p_0.
- $\overrightarrow{EE}$: By duality, we have to show $\|\nabla \varphi(L^*) f\|_{p_1} \lesssim \|\nabla f\|_{p_1}$. This is the same as $\overrightarrow{BB}$ by changing L to L^* and p_0 to p_1.
- $\overrightarrow{FF}$: By duality, this is the same as $\overrightarrow{AA}$ by changing L to L^* and q_0 to q_1.

46   5. RIESZ TRANSFORMS AND FUNCTIONAL CALCULI

Step 2. Maximal horizontal arrows.

• $\overrightarrow{BA}$: As for $\overrightarrow{AA}$, we begin with
$$\|\nabla\varphi(L)f\|_{q_0} \lesssim \|L^{1/2}\varphi(L)f\|_{q_0} \sim \|\varphi(L)L^{1/2}f\|_{q_0}.$$
Next, we continue with
$$\|\varphi(L)L^{1/2}f\|_{q_0} \lesssim \|z^{\alpha(B,A)}\varphi\|_\infty \|L^{-\alpha(B,A)}L^{1/2}f\|_{q_0}$$
by Theorem 5.1, then
$$\|L^{-\alpha(B,A)}L^{1/2}f\|_{q_0} \lesssim \|L^{1/2}f\|_{p_0} \lesssim \|\nabla f\|_{p_0},$$
by the Hardy-Littlewood-Sobolev inequality for L and the definition of $\alpha(B,A)$, and the reverse inequality at p_0.

• $\overrightarrow{DC}$: Since $p_-(L) < p_0 < (p_1)' < p_+(L)$, by Proposition 5.3 and Theorem 5.1,
$$\|\varphi(L)f\|_{(p_1)'} \lesssim \|z^{\alpha(D,C)}\varphi\|_\infty \|L^{-\alpha(D,C)}f\|_{(p_1)'} \lesssim \|f\|_{p_0}.$$

• $\overrightarrow{FE}$: By duality, this is the same as $\overrightarrow{BA}$ by changing L to L^*, p_0 to p_1 and q_0 to q_1.

Step 3. Maximal vertical arrows.

Recall that G and H are points in $\mathcal{P}$ given by $G = (\frac{1}{p_0}, -1)$ and $H = (\frac{1}{(q_1)'}, 1)$.

• $\overrightarrow{DB}$: Using that the Riesz transform is L^{p_0} bounded and Theorem 5.1, one has
$$\|\nabla\varphi(L)f\|_{p_0} \lesssim \|L^{1/2}\varphi(L)f\|_{p_0} \lesssim \|z^{1/2}\varphi\|_\infty \|f\|_{p_0}.$$

• $\overrightarrow{BD}$: We have
$$\|\varphi(L)f\|_{p_0} \lesssim \|L^{-1/2}\varphi(L)L^{1/2}f\|_{p_0} \lesssim \|z^{-1/2}\varphi\|_\infty \|L^{1/2}f\|_{p_0} \lesssim \|\nabla f\|_{p_0}.$$

• $\overrightarrow{FH}$: Since $(q_1)' > p_-(L)$, one has the $L^{(q_1)'}$ boundedness of the Riesz transform associated to L:
$$\|\nabla\varphi(L)\operatorname{div} f\|_{(q_1)'} \lesssim \|L^{1/2}\varphi(L)\operatorname{div} f\|_{(q_1)'}.$$
Then writing $L^{1/2}\varphi(L) = L\varphi(L)L^{-1/2}$ and using Theorem 5.1 yield
$$\|L^{1/2}\varphi(L)\operatorname{div} f\|_{(q_1)'} \lesssim \|z\varphi\|_\infty \|L^{-1/2}\operatorname{div} f\|_{(q_1)'}.$$
Next, using that the Riesz transform associated to L^* is bounded on L^{q_1}, one concludes by
$$\|L^{-1/2}\operatorname{div} f\|_{(q_1)'} \lesssim \|f\|_{(q_1)'}.$$

• $\overrightarrow{HF}$: Since $(q_1)' > p_-(L)$, $L^{1/2}$ is bounded from $L^{(q_1)'}$ to $\dot{W}^{-1,(q_1)'}$, hence
$$\|\varphi(L)f\|_{\dot{W}^{-1,(q_1)'}} = \|L^{1/2}\varphi(L)L^{-1/2}f\|_{\dot{W}^{-1,(q_1)'}} \lesssim \|\varphi(L)L^{-1/2}f\|_{(q_1)'}.$$
Next, by Theorem 5.1,
$$\|\varphi(L)L^{-1/2}f\|_{(q_1)'} \lesssim \|z^{-1}\varphi\|_\infty \|L^{1/2}f\|_{(q_1)'}$$
We finish with $L^{1/2}$ bounded from $\dot{W}^{(q_1)'}$ to $L^{(q_1)'}$ since $(q_1)' > p_-(L)$.

• $\overrightarrow{GA}$: Same as $\overrightarrow{FH}$ by changing $(q_1)'$ to q_0.
• $\overrightarrow{AG}$: Same as $\overrightarrow{HF}$ by changing $(q_1)'$ to q_0.
• $\overrightarrow{EC}$: By duality, this is the same as $\overrightarrow{DB}$ by changing L to L^* and p_0 to p_1.
• $\overrightarrow{CE}$: By duality, this is the same as $\overrightarrow{BD}$ by changing L to L^* and p_0 to p_1. □

We conclude this section with the following corollary.

COROLLARY 5.6. *If $q_-(L) < p < q_+(L)$, then L has a bounded holomorphic functional calculus on $\dot{W}^{1,p}$.* [4]

PROOF. Since $\overrightarrow{(q_-(L), q_+(L))} = \mathcal{I}(L)$, this functional calculus corresponds to minimal arrow $\overrightarrow{MM}$ for $M = (\frac{1}{p}, 1)$. □

5.4. More on the Kato diagram

As we have seen, we can move vertically up and down in $\mathcal{P}$. However, one can authorize more downward arrows. Doing this and modifying slightly the definition of the numbers $r_\pm(L)$ in Theorem 4.9, we shall show that $r_+(L) = p_+(L)$ and obtain a lower bound on $r_-(L)$.

First we claim that the proof given for the reverse inequalities $\|L^{1/2}f\|_p \lesssim \|\nabla f\|_p$ extends as follows.

PROPOSITION 5.7. *For all $\mu \in (\omega, \pi)$ and all $\varphi \in H^\infty(\Sigma_\mu)$,*

$$\|L^{1/2}\varphi(L)f\|_p \lesssim \|\varphi\|_\infty \|\nabla f\|_p$$

whenever $\tilde{p}_-(L) = r_-(L) < p < p_+(L)$.

PROOF. The case $p > 2$ is a simple consequence of this inequality for $L^{1/2}$ together with the bounded holomorphic functional calculus on L^p for $p < p_+(L)$.

We now turn to $p < 2$. We begin with the representation formula (2.6) for $\psi(z) = z^{1/2}\varphi(z)$. One has

$$\psi(L) = \int_{\Gamma_+} e^{-zL}\eta_+(z)\,dz + \int_{\Gamma_-} e^{-zL}\eta_-(z)\,dz$$

and

$$\eta_\pm(z) = \frac{1}{2\pi i}\int_{\gamma_\pm} e^{\zeta z}\psi(\zeta)\,d\zeta, \quad z \in \Gamma_\pm,$$

valid provided the technical assumption $|\varphi(\zeta)| \leq C(1+|\zeta|)^{-1/2-s}$ for some $s > 0$. If one defines the primitive $N_\pm(z)$ of respectively $\eta_\pm(z)$ which vanishes at infinity then, under the technical assumption, one may integrate by parts and, since the terms at 0 cancel each other, one finds

$$\psi(L) = \int_{\Gamma_+} Le^{-zL}N_+(z)\,dz + \int_{\Gamma_-} Le^{-zL}N_-(z)\,dz.$$

Furthermore, one has

$$|N_\pm(z)| \leq C|z|^{-1/2}\|\varphi\|_\infty, \quad z \in \Gamma_\pm.$$

Hence, by repeating the argument for $L^{1/2}$ where integration on the positive axis is replaced by integration on half-rays $\Gamma_\pm$, we obtain

$$\|\psi(L)f\|_p \lesssim \|\varphi\|_\infty \|\nabla f\|_p$$

whenever $\tilde{p}_-(L) < p < 2$ and φ sastisfies the technical condition, which is removed by a limiting argument. □

Let us discuss the sharpness of the bounds obtained above.

---

[4] We do not know if the converse holds.

PROPOSITION 5.8. *Assume that for all $\mu \in (\omega, \pi)$ and all $\varphi \in H^\infty(\Sigma_\mu)$,*
$$\|L^{1/2}\varphi(L)f\|_p \lesssim \|\varphi\|_\infty \|\nabla f\|_p.$$
Then $\sup(\overline{p_-}(L), 1) \leq p \leq p_+(L)$ *where $\overline{p_-}(L)$ is defined as follows: the point $\overline{B} = (\frac{1}{\overline{p_-}(L)}, 1)$ is symmetric to $F = (\frac{1}{(q_+(L^*))'}, -1)$ with respect to $D = (\frac{1}{(p_+(L^*))'}, 0) = (\frac{1}{p_-(L)}, 0)$.*

One can see that $\overline{p_-}(L) \leq \tilde{p}_-(L)$. This lower bound $\overline{p_-}(L)$ is the best one can obtain by a convexity method. If $n \leq 4$, $\tilde{p}_-(L) = 1$ so there is nothing more to say. In dimensions $n \geq 5$, the lower bound is optimal for any operator L for which this inequality cannot be improved, which is the same as saying that $p_+(L^*) \geq (q_+(L^*))^*$ is optimal (See Section 3.5 for this).

PROOF. Fix $\varphi \in H^\infty(\Sigma_\mu)$ with $\|\varphi\|_\infty = 1$. We have for all $t \in \mathbb{R}$ and some $0 < c < \pi/2$,
$$\|L^{1/2+it}\varphi(L)f\|_p \lesssim e^{c|t|}\|\nabla f\|_p = e^{c|t|}\|f\|_{\dot{W}^{1,p}}.$$
We also have from the HLSK diagram that for all $(q_+(L^*))' < q < p_+(L)$ and $t \in \mathbb{R}$,
$$\|L^{-1/2+it}\varphi(L)f\|_q \lesssim e^{c|t|}\|f\|_{\dot{W}^{-1,q}}.$$
By complex interpolation, we obtain
$$\|\varphi(L)f\|_r \lesssim \|f\|_r$$
for $1/r$ the middle of $1/p$ and $1/q$. By Theorem 5.1, it is necessary that $p_-(L) \leq r \leq p_+(L)$.

Choosing q arbitrarily close to $p_+(L)$ forces $p \leq p_+(L)$. Choosing q arbitrarily close to $(q_+(L^*))'$ forces $p \geq \overline{p_-}(L)$ given the definition of this number. $\square$

Geometrically, this provides us with a family of authorized downarrows that may not be contained in $\mathcal{P}$. We assume that $\tilde{p}_0 = (p_0)_* > 1$ otherwise we exclude from this discussion points $\{(\frac{1}{p}, s)\}$ with $\frac{1}{p} \geq 1$. Consider the closed convex polygon $\mathcal{P}_{in} = A\tilde{B}DFEC$ in the plane $\{(\frac{1}{p}, s)\}$ with A, D, F, E, C as before and $\tilde{B} = (\frac{1}{\tilde{p}_0}, 1)$ and also the closed convex polygon $\mathcal{P}_{ext} = \tilde{A}\tilde{B}\tilde{D}FE$, where $\tilde{A} = (\frac{1}{(p_1)'}, 1)$ and $\tilde{D} = (\frac{1}{\tilde{p}_0}, 0)$. See Figure 2.

PROPOSITION 5.9. *For any downarrow $\overrightarrow{MN}$, $M = (\frac{1}{p}, s)$ $N = (\frac{1}{p}, \sigma)$ with one extremity $\mathcal{P}_{in}$ and the other in $\mathcal{P}_{ext}$ we have for all $\mu \in (\omega, \pi)$ and all $\varphi \in H^\infty(\Sigma_\mu)$,*
$$\|\varphi(L)f\|_{\dot{W}^{\sigma,p}} \lesssim \|z^{\alpha(M,N)}\varphi\|_{L^\infty(\Sigma_\mu)}\|f\|_{\dot{W}^{s,p}},$$
provided the quantity $\|z^{\alpha(M,N)}\varphi\|_{L^\infty(\Sigma_\mu)}$ is finite. The implicit constant depends on $\mu, \omega, p, q, s, \sigma$.

PROOF. The previous proposition shows that for the maximal downarrows $\overrightarrow{AC}$ and $\overrightarrow{\tilde{B}D}$ the corresponding inequality of the statement are valid. Combining this with all possible vertical authorized arrows in $\mathcal{P}$, the convexity property of complex interpolation gives us the desired result. $\square$

Of course, the downarrows can be reversed exactly when M and N belong to the HLSK diagram $\mathcal{P}$.

The expected maximal convex set $\mathcal{P}_{in}$ would be with $\tilde{B}$ replaced by $\overline{B}$ where $\overline{B}$ is the symmetric point of F with respect to D. But our arguments do not suffice

5.4. MORE ON THE KATO DIAGRAM

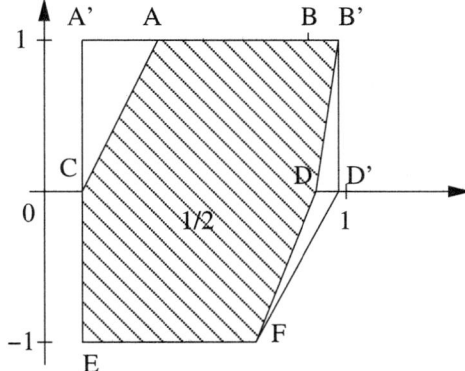

FIGURE 2. The convex polygons $\mathcal{P}_{in}$ (shaded) and $\mathcal{P}_{ext}$. The points A', B', D' correspond to $\tilde{A}, \tilde{B}, \tilde{D}$ in the text. Downarrows MN with $M \in \mathcal{P}_{in}$ and $N \in \mathcal{P}_{ext}$ or $M \in \mathcal{P}_{ext}$ and $N \in \mathcal{P}_{in}$ are authorized.

to prove the inequality corresponding to the vertical downarrow $\overrightarrow{\overline{BD}}$ with $\overline{D}$ the vertical projection of $\overline{B}$ on the $\frac{1}{p}$ axis.

CHAPTER 6

Square function estimates

In this chapter, we study vertical square functions of two different types which are representative of larger classes of square functions. Then, we prove some weak type and strong type inequalities for non-tangential square functions.

6.1. Necessary and sufficient conditions for boundedness of vertical square functions

Define the quadratic functionals for $f \in L^2$

$$g_L(f)(x) = \left(\int_0^\infty |(L^{1/2}e^{-tL}f)(x)|^2 \, dt\right)^{1/2}$$

and

$$G_L(f)(x) = \left(\int_0^\infty |(\nabla e^{-tL}f)(x)|^2 \, dt\right)^{1/2}.$$

The L^2 theory of quadratic estimates for operators having a bounded holomorphic functional calculus on L^2 implies

(6.1) $$\|g_L(f)\|_2 \sim \|f\|_2.$$

In fact, such an inequality is equivalent to the bounded holomorphic functional calculus on L^2. Moreover, the family $((tL)^{1/2}e^{-tL})_{t>0}$ in g_L can be replaced by more general functions of L. [1]

As for G_L,

(6.2) $$\|G_L(f)\|_2 \sim \|f\|_2$$

is a consequence of ellipticity and

$$\int_{\mathbb{R}^n} |f(x)|^2 \, dx = 2\Re \iint_{\mathbb{R}^n \times (0,\infty)} A(x)(\nabla e^{-tL}f)(x) \cdot \overline{(\nabla e^{-tL}f)(x)} \, dxdt.$$

This equality is obtained as follows. We have

$$\|f\|_2^2 = -\int_0^\infty \frac{d}{dt}\|e^{-tL}f\|_2^2 \, dt$$

$$= 2\Re \iint_{\mathbb{R}^n \times (0,\infty)} (Le^{-tL}f)(x)\overline{(e^{-tL}f)(x)} \, dxdt$$

and it remains to integrate by parts in the x variable using the definition of L.

We are interested in the L^p counterparts of these results. Define

$$\mathcal{S}(L) = \{1 < p < \infty; \, \forall f \in L^2 \cap L^p \quad \|g_L(f)\|_p \sim \|f\|_p.\}$$

---

[1] All this is due to McIntosh and Yagi, [54] and [67].

and $\mathcal{S}_-(L) = \mathcal{S}(L) \cap (1,2)$ and
$$\mathcal{G}(L) = \{1 < p < \infty;\ \forall f \in L^2 \cap L^p \quad \|G_L(f)\|_p \sim \|f\|_p.\},$$
$\mathcal{G}_-(L) = \mathcal{G}(L) \cap (1,2)$ and $\mathcal{G}_+(L) = \mathcal{G}(L) \cap (2,\infty)$. Recall $\operatorname{Int} \mathcal{J}(L) = (p_-(L), p_+(L))$ and $\operatorname{Int} \mathcal{N}(L) = (q_-(L), q_+(L))$.

THEOREM 6.1. (1) *The interior of $\mathcal{S}(L)$ is $(p_-(L), p_+(L))$.* [2]
(2) *The interior of $\mathcal{G}(L)$ is $(q_-(L), q_+(L))$.* [3]

Roughly, this theorem says that, up to endpoints, the range of $p \in (1,\infty)$ for which g_L defines a new norm on L^p is the same as the one of boundedness of the semigroup and that the range of $p \in (1,\infty)$ for which G_L defines a new norm on L^p is the same as the one of boundedness of $(\sqrt{t}\,\nabla e^{-tL})_{t>0}$.

Due to Theorems 4.2 and 4.6, the connections to intervals of exponents p for which one has L^p boundedness for the Riesz transform $\nabla L^{-1/2}$ is as follows.

COROLLARY 6.2. (1) *The intervals $\mathcal{I}_-(L)$, $\mathcal{S}_-(L)$ and $\mathcal{G}_-(L)$ have same interiors.*
(2) *The intervals $\mathcal{I}_+(L)$ and $\mathcal{G}_+(L)$ have same interiors.*

Hence, there is again a dichotomy $p < 2$ vs $p > 2$ in the description. We turn to the proof of the theorem.

PROOF OF THEOREM 6.1. By Proposition 3.2, we may freely replace $\mathcal{J}_\pm(L)$ by one of the intervals $\mathcal{K}_\pm(L)$, $\mathcal{M}_\pm(L)$. The argument has several steps:

Step 1: $p_0 > 2$ and $(\sqrt{t}\,\nabla e^{-tL})_{t>0}$ satisfies $L^2 - L^{p_0}$ off-diagonal estimates imply $\|G_L(f)\|_p \lesssim \|f\|_p$ for $2 < p < p_0$.

Step 2: $p_0 > 2$ and $(e^{-tL})_{t>0}$ satisfies $L^2 - L^{p_0}$ off-diagonal estimates imply $\|g_L(f)\|_p \lesssim \|f\|_p$ for $2 < p < p_0$.

Step 3: $p_0 < 2$ and $(e^{-tL})_{t>0}$ satisfies $L^{p_0} - L^2$ off-diagonal estimates imply $\|g_L(f)\|_p \lesssim \|f\|_p$ for $p_0 < p < 2$.

Step 4: $p_0 < 2$ and $(\sqrt{t}\,\nabla e^{-tL})_{t>0}$ satisfies $L^{p_0} - L^2$ off-diagonal estimates imply $\|G_L(f)\|_p \lesssim \|f\|_p$ for $p_0 < p < 2$.

Step 5: Reverse L^p inequality for g_L when $p_-(L) < p < p_+(L)$.

Step 6: $\|g_L(f)\|_p \sim \|f\|_p$ implies $(e^{-tL})_{t>0}$ L^p bounded.

Step 7: $\|G_L(f)\|_p \lesssim \|f\|_p$ implies $(\sqrt{t}\,\nabla e^{-tL})_{t>0}$ L^q-bounded for q in the interval between 2 and p.

Step 8: Reverse L^p inequality for G_L when $1 < p < \infty$.

The combination of steps 2, 3 and 5 shows that $\operatorname{Int} \mathcal{J}(L)$ is contained in $\mathcal{S}(L)$. Step 6 implies that $\mathcal{S}(L)$ is contained in $\mathcal{J}(L)$.

The combination of steps 1 and 4 shows that $\operatorname{Int} \mathcal{N}(L)$ is contained in $\operatorname{Int} \mathcal{G}(L)$. Step 7 and Step 8 show the converse.

---

[2] We develop an argument based on the extension of Calderón-Zygmund theory in Chapter 1. However, there is an other argument as follows: we have shown that the interior of $\mathcal{J}(L)$ is the maximal open interval of exponents p for which L has a bounded holomorphic functional calculus of L^p. Take such a p. Then, apply Le Merdy's Theorem 3 in [48] to conclude that g_L defines a new norm on L^p. More is proved in [48]: the operators $(tL)^{1/2}e^{-tL}$ in the definition of g_L can be replaced by more general functions of L. We choose not to develop this here.

[3] The statement is valid for square functions where the semigroup in the definition of G_L is replaced by more general functions of L. Again, we do not go into such considerations.

6.1. VERTICAL SQUARE FUNCTIONS

Proof of Step 1: $p_0 > 2$ and $(\sqrt{t}\nabla e^{-tL})_{t>0}$ satisfies $L^2 - L^{p_0}$ off-diagonal estimates implies $\|G_L(f)\|_p \lesssim \|f\|_p$ for $2 < p < p_0$. We apply Theorem 1.2 to the sublinear operator $T = G_L$. We choose as usual $A_r = I - (I - e^{-r^2 L})^m$ for m a large enough integral number.

We have to check (1.4). Let B be a ball and $r = r(B)$ be its radius and $k = 1, \ldots, m$. By Minkowski integral inequality and $p_0 > 2$, we have

$$\left(\frac{1}{|B|}\int_B |G_L(e^{-kr^2 L}f)|^{p_0}\right)^{2/p_0} \leq \int_0^\infty \left(\frac{1}{|B|}\int_B |\nabla e^{-tL}(e^{-kr^2 L}f)|^{p_0}\right)^{2/p_0} dt$$

Using the hypothesis of p_0, the commutativity property of the semigroup and applying the scalar inequality (4.6) to $e^{-tL}f$ for each $t > 0$, we have

$$\left(\frac{1}{|B|}\int_B |\nabla e^{-tL}(e^{-kr^2 L}f)|^{p_0}\right)^{1/p_0} \leq \sum_{j \geq 1} g(j) \left(\frac{1}{|2^{j+1}B|}\int_{2^{j+1}B} |\nabla e^{-tL}f|^2\right)^{1/2}$$

Squaring this and using that $\sum g(j) < \infty$, we obtain

$$\int_0^\infty \left(\frac{1}{|B|}\int_B |\nabla e^{-tL}(e^{-kr^2 L}f)|^{p_0}\right)^{2/p_0} dt \leq C \int_0^\infty \sum_{j \geq 1} \frac{g(j)}{|2^{j+1}B|}\int_{2^{j+1}B} |\nabla e^{-tL}f|^2 \, dt$$

Exchanging the sum and the integral, the latter is equal to

$$C \sum_{j \geq 1} g(j) \frac{1}{|2^{j+1}B|} \int_{2^{j+1}B} |G_L(f)|^2$$

which is controlled by $CM(G_L(f)^2)(y)$ for any $y \in B$.

Next, we establish (1.3). Let B be a ball and $r = r(B)$ be its radius. Let $f \in L^2$. Decompose $f = f_1 + f_2 + f_3 \ldots$ where $f_j = f\chi_{C_j}$ and C_j are defined as usual. We start from

$$\left(\frac{1}{|B|}\int_B |G_L((I - e^{-r^2 L})^m f)|^2\right)^{1/2} \leq \sum_{j \geq 1} \left(\frac{1}{|B|}\int_B |G_L((I - e^{-r^2 L})^m f_j)|^2\right)^{1/2}.$$

For $j = 1$, we merely use the L^2 boundedness of G_L in (6.2) and that of $(I - e^{-r^2 L})^m$ to obtain

$$\left(\frac{1}{|B|}\int_B |G_L((I - e^{-r^2 L})^m f_1)|^2\right)^{1/2} \leq C \left(\frac{1}{|4B|}\int_{4B} |f|^2\right)^{1/2}.$$

For $j \geq 2$, we write

$$\frac{1}{|B|}\int_B |G_L((I - e^{-r^2 L})^m f_j)|^2 = \int_0^\infty \frac{1}{|B|}\int_B |\sqrt{t}\nabla(I - e^{-kr^2 L})^m e^{-tL} f_j|^2 \, dx \frac{dt}{t}.$$

and we use the estimates in the second argument of the proof of Theorem 4.6 to obtain a bound

$$\frac{C 2^{nj}}{|2^{j+1}B|}\int_{2^{j+1}B} |f|^2 \left(\int_0^\infty \frac{1}{4^{jm}} \inf\left(\left(\frac{t}{4^j r^2}\right)^{1/2}, \left(\frac{4^j r^2}{t}\right)^{m-1/2}\right) \frac{dt}{t}\right)^2,$$

hence as $m \geq 1$,

$$C 2^{nj} 2^{-4mj} \frac{1}{|2^{j+1}B|}\int_{2^{j+1}B} |f|^2.$$

Choosing further $m > n/4$ allows to sum in $j \geq 2$ and to conclude for (1.3).

Proof of Step 2: $p_0 > 2$ and $(e^{-tL})_{t>0}$ satisfies $L^2 - L^{p_0}$ off-diagonal estimates imply $\|g_L(f)\|_p \lesssim \|f\|_p$ for $2 < p < p_0$. We apply Theorem 1.2 to the sublinear operator $T = g_L$. We choose as usual $A_r = I - (I - e^{-r^2 L})^m$ for m a large enough integral number.

We have to check (1.4). Let B be a ball and $r = r(B)$ be its radius and $k = 1, \ldots, m$. By Minkowski integral inequality and $p_0 > 2$, we have

$$\left(\frac{1}{|B|} \int_B |g_L(e^{-kr^2 L} f)|^{p_0}\right)^{2/p_0} \leq \int_0^\infty \left(\frac{1}{|B|} \int_B |(tL)^{1/2} e^{-tL}(e^{-kr^2 L} f)|^{p_0}\right)^{2/p_0} \frac{dt}{t}.$$

Using the hypothesis of p_0, and following already used arguments, we have that

$$\left(\frac{1}{|B|} \int_B |(e^{-kr^2 L} g)|^{p_0}\right)^{2/p_0} \leq \sum_{j \geq 1} \frac{c_j}{|2^{j+1} B|} \int_{2^{j+1} B} |g|^2$$

with $c_j = C e^{-c 4^j}$ for some positive constants c, C. Applying the commutativity property of the semigroup and this inequality to $g = (tL)^{1/2} e^{-tL} f$ for each $t > 0$, we have

$$\left(\frac{1}{|B|} \int_B |g_L(e^{-kr^2 L} f)|^{p_0}\right)^{2/p_0} \leq \int_0^\infty \left\{\sum_{j \geq 1} \frac{c_j}{|2^{j+1} B|} \int_{2^{j+1} B} |(tL)^{1/2} e^{-tL} f|^2\right\} \frac{dt}{t}.$$

As the latter expression equals

$$\sum_{j \geq 1} \frac{c_j}{|2^{j+1} B|} \int_{2^{j+1} B} |g_L(f)|^2$$

we obtain a bound in $CM(g_L(f)^2)(y)$ for any $y \in B$ as desired.

Next, we establish (4.5). Let B be a ball and $r = r(B)$ be its radius. Let $f \in L^2$. Decompose $f = f_1 + f_2 + f_3 \ldots$ where $f_j = f \chi_{C_j}$ and $C_j = C_j(B)$ are defined as usual. We start from

$$\left(\frac{1}{|B|} \int_B |g_L((I - e^{-kr^2 L})^m f)|^2\right)^{1/2} \leq \sum_{j \geq 1} \left(\frac{1}{|B|} \int_B |g_L((I - e^{-kr^2 L})^m f_j)|^2\right)^{1/2}.$$

For $j = 1$, we merely use the L^2 boundedness of g_L in (6.1) and that of $(I - e^{r^2 L})^m$ to obtain

$$\left(\frac{1}{|B|} \int_B |g_L((I - e^{-kr^2 L})^m f_1)|^2\right)^{1/2} \leq C \left(\frac{1}{|4B|} \int_{4B} |f|^2\right)^{1/2}.$$

For $j \geq 2$, we write

$$\frac{1}{|B|} \int_B |g_L((I - e^{-kr^2 L})^m f_j)|^2 = \int_0^\infty \frac{1}{|B|} \int_B |(tL)^{1/2} e^{-tL}(I - e^{-kr^2 L})^m f_j|^2 \, dx \frac{dt}{t}.$$

As in Subsection 4.1.1, one may use the representation (2.6) with the function $\varphi(z) = (tz)^{1/2} e^{-tz} (1 - e^{r^2 z})^m$. The corresponding functions $\eta_\pm$ satisfy the estimates

$$(6.3) \qquad |\eta_\pm(z)| \leq \frac{C t^{1/2}}{(|z| + t)^{3/2}} \inf\left(1, \frac{r^{2m}}{(|z| + t)^m}\right), \quad z \in \Gamma_\pm.$$

Since for any $0 < \beta < \frac{\pi}{2} - \omega$, $(e^{-zL})_{z \in \Sigma_\beta}$ satisfies L^2 off-diagonal estimates, using (2.6) and the above estimate for $\eta_\pm$, $\|(tL)^{1/2} e^{-tL}(I - e^{-kr^2 L})^m f_j\|_{L^2(B)}$ is bounded by

$$C \int_{\Gamma_+} e^{-\frac{c4^j r^2}{|z|}} \frac{t^{1/2}}{(|z|+t)^{3/2}} \frac{r^{2m}}{(|z|+t)^m} |dz| \, \|f\|_{L^2(C_j)}$$

plus the similar term corresponding to integration on Γ_-. Using (4.4), this gives us the bound

$$\frac{C}{4^{jm}} \inf\left(\left(\frac{t}{4^j r^2}\right)^{1/2}, \left(\frac{4^j r^2}{t}\right)^m \right) \|f\|_{L^2(C_j)}.$$

Squaring and integrating with respect to t, we find

$$\frac{1}{|B|} \int_B |g_L((I - e^{-kr^2 L})^m f_j)|^2 \leq C 2^{jn} 4^{-mj} \frac{1}{|2^{j+1} B|} \int_{2^{j+1} B} |f|^2$$

and this readily implies (4.5).

Proof of Step 3: $p_0 < 2$ and $(e^{-tL})_{t>0}$ satisfies $L^{p_0} - L^2$ off-diagonal estimates imply $\|g_L(f)\|_p \lesssim \|f\|_p$ for $p_0 < p < 2$. We apply Theorem 1.1 to $T = g_L$. We choose as usual $A_r = I - (I - e^{-r^2 L})^m$ for m a large enough integral number.

Equation (1.2) is a direct consequence of the assumption on the semigroup. We then turn to the verification of (1.1). Let B be a ball and $r = r(B)$ its radius and $j \geq 2$. Let f be a function supported in B. We have

$$\frac{1}{|2^{j+1} B|} \int_{C_j} |g_L((I - e^{-kr^2 L})^m f)|^2$$

$$= \int_0^\infty \frac{1}{|2^{j+1} B|} \int_{C_j} |(tL)^{1/2} e^{-tL} (I - e^{-kr^2 L})^m f|^2 \, dx \frac{dt}{t}.$$

Using similar arguments as in the previous step and using the hypothesis on the semigroup, $\|(tL)^{1/2} e^{-tL} (I - e^{-r^2 L})^m f\|_{L^2(C_j)}$ is bounded by

$$C \int_{\Gamma_+} e^{-\frac{c4^j r^2}{|z|}} \frac{1}{|z|^{\gamma/2}} \frac{t^{1/2}}{(|z|+t)^{3/2}} \frac{r^{2m}}{(|z|+t)^m} |dz| \, \|f\|_{L^{p_0}(B)}$$

plus the similar term corresponding to integration on Γ_-. Here $\gamma = \gamma_{p_0} = |\frac{n}{2} - \frac{n}{p_0}|$. Using (4.4), this gives us the bound

$$\frac{C}{4^{jm}(2^j r)^\gamma} \inf\left(\left(\frac{t}{4^j r^2}\right)^{1/2}, \left(\frac{4^j r^2}{t}\right)^m \right) \|f\|_{L^{p_0}(B)}.$$

Squaring and integrating with respect to t, we find

$$\left(\frac{1}{|2^{j+1} B|} \int_{C_j} |g_L((I - e^{-r^2 L})^m f)|^2\right)^{1/2} \leq C 2^{-jn/2} 4^{-mj} 2^{-j\gamma} \left(\frac{1}{|B|} \int_B |f|^{p_0}\right)^{1/p_0}$$

and this readily implies (1.1) and m is chosen with $2m + \gamma > n/2$.

Proof of Step 4: $p_0 < 2$ and $(\sqrt{t} \nabla e^{-tL})_{t>0}$ satisfies $L^{p_0} - L^2$ off-diagonal estimates imply $\|G_L(f)\|_p \lesssim \|f\|_p$ for $p_0 < p < 2$. We apply Theorem 1.1 to $T = G_L$. We choose as usual $A_r = I - (I - e^{-r^2 L})^m$ for m a large enough integral number.

Equation (1.2) is a direct consequence of the assumption on the semigroup. We then turn to the verification of (1.1). Let B be a ball and $r = r(B)$ its radius and $j \geq 2$. Let f be a function supported in B. We have

$$\frac{1}{|2^{j+1}B|} \int_{C_j} |G_L((I - e^{-r^2 L})^m f)|^2$$
$$= \int_0^\infty \frac{1}{|2^{j+1}B|} \int_{C_j} |\sqrt{t} \nabla e^{-tL}(I - e^{-kr^2 L})^m f|^2 \, dx \frac{dt}{t}.$$

Proposition 3.16 shows that we may replace the hypothesis by $(\sqrt{z}\nabla e^{-zL})_{z \in \Sigma_\beta}$ satisfies $L^{p_0} - L^2$ off-diagonal estimates for any $\beta \in (0, \frac{\pi}{2} - \omega)$ (up to changing p_0 to an arbitrary larger value). Using then similar arguments as in the step 1, $\|\sqrt{t}\nabla e^{-tL}(I - e^{-kr^2 L})^m f\|_{L^2(C_j)}$ is bounded by

$$C \int_{\Gamma_+} e^{-\frac{c 4^j r^2}{|z|}} \frac{1}{|z|^{\gamma/2 + 1/2}} \frac{t^{1/2}}{(|z| + t)} \frac{r^{2m}}{(|z| + t)^m} |dz| \, \|f\|_{L^{p_0}(B)}$$

plus the similar term corresponding to integration on Γ_-. Here $\gamma = \gamma_{p_0} = |\frac{n}{2} - \frac{n}{p_0}|$. Using (4.4), this gives us the bound

$$\frac{C}{4^{jm}(2^j r)^\gamma} \inf\left(\left(\frac{t}{4^j r^2}\right)^{1/2}, \left(\frac{4^j r^2}{t}\right)^{m - 1/2} \right) \|f\|_{L^{p_0}(B)}.$$

Squaring and integrating with respect to t, we find

$$\left(\frac{1}{|2^{j+1}B|} \int_{C_j} |G_L((I - e^{-kr^2 L})^m f)|^2 \right)^{1/2} \leq C 2^{-jn/2} 4^{-mj} 2^{-j\gamma} \left(\frac{1}{|B|} \int_B |f|^{p_0} \right)^{1/p_0}$$

and this readily implies (1.1) if m is chosen with $2m + \gamma > n/2$.

Proof of Step 5: Reverse L^p inequality for g_L when $p_-(L) < p < p_+(L)$. By functional calculus for L, we have the representation formula for the identity on L^2

(6.4) $$f = 2 \int_0^\infty tLe^{-2tL} f \frac{dt}{t}.$$

Hence, dualizing against g in L^2, writing

$$\langle tLe^{-2tL} f, g \rangle = \langle (tL)^{1/2} e^{-tL} f, (tL^*)^{1/2} e^{-tL^*} g \rangle$$

and using Cauchy-Schwarz inequality yield

$$\left| \int_{\mathbb{R}^n} f\bar{g} \right| \leq \int_{\mathbb{R}^n} g_L(f) g_{L^*}(g).$$

Recall that $\operatorname{Int} \mathcal{J}(L) = (p_-(L), p_+(L))$ is the largest open interval of exponents p in $(1, \infty)$ for which the semigroup $(e^{-tL})_{t>0}$ is L^p bounded. Then, by steps 2 and 3, g_L is L^p bounded for $p \in \operatorname{Int} \mathcal{J}(L)$. Let $p \in \operatorname{Int} \mathcal{J}(L)$. Applying this to L^* since $p' \in \operatorname{Int} \mathcal{J}(L^*)$, we have

$$\|g_{L^*}(g)\|_{p'} \leq C \|g\|_{p'},$$

hence,

$$\left| \int_{\mathbb{R}^n} f\bar{g} \right| \leq C \|g_L(f)\|_p \|g\|_{p'}.$$

In conclusion,

$$\|f\|_p \leq C \|g_L(f)\|_p$$

6.1. VERTICAL SQUARE FUNCTIONS

for $p \in \text{Int } \mathcal{J}(L)$.

Proof of Step 6: $\|g_L(f)\|_p \sim \|f\|_p$ implies $(e^{-tL})_{t>0}$ L^p bounded. It follows easily form the definition of g_L and the commutativity properties of the semigroup that for all $s > 0$,

$$g_L(e^{-sL}f) \leq g_L(f)$$

in the pointwise sense, hence in L^p norm. In particular, this and the hypothesis yield $\|e^{-sL}f\|_p \leq C\|f\|_p$ uniformly in $s > 0$.

Proof of Step 7: $\|G_L(f)\|_p \lesssim \|f\|_p$ implies $(\sqrt{t}\nabla e^{-tL})_{t>0}$ L^q-bounded for q between 2 and p. [4]

The argument of step 5 does not apply for the semigroup and the gradient operator do not commute. We rely instead on some trick using complex interpolation. Define

$$\widetilde{G}_L(f)(x) = \left(\int_0^\infty |(\nabla(tL)e^{-tL}f)(x)|^2 \, dt\right)^{1/2}.$$

We assume $p > 2$. The argument for $p < 2$ is entirely similar. We show that $\|G_L(f)\|_p \lesssim \|f\|_p$ implies $\|\widetilde{G}_L(f)\|_q \lesssim \|f\|_q$ for $2 < q < p$. Assume this is done. We prove the L^q boundedness of $\sqrt{s}\nabla e^{-sL}$. Without loss of generality, we may asume $s = 1$. Write

$$|\nabla e^{-L}f| \leq \int_1^2 |\nabla e^{-sL}f| \, ds + \int_1^2 |\nabla e^{-L}f - \nabla e^{-sL}f| \, ds$$

$$\leq \int_1^2 |\nabla e^{-sL}f| \, ds + \int_1^2 \left|\int_1^s \nabla L e^{-tL}f \, dt\right| ds$$

$$\leq \int_1^2 |\nabla e^{-sL}f| \, ds + \int_1^2 |\nabla(tL)e^{-tL}f| \, dt.$$

Hence,

$$|\nabla e^{-L}f| \leq G_L(f) + \widetilde{G}_L(f).$$

and the L^q boundedness of ∇e^{-L} follows from that of G_L and $\widetilde{G}_L$.

It remains to prove the L^q boundedness of the latter. To this end, we follow the proof of Stein's complex interpolation theorem after dualizing. Fix q with $2 < q < p$.

Let f be a $\mathbb{C}$-valued simple function with $\|f\|_q = 1$ and $g = (g_t)_{t>0}$ be a H-valued simple function where $H = L^2((0,\infty),\mathbb{C}^n;dt)$ with $\|g\|_{q'} = 1$. Write $f(x) = \sum_k a_k \chi_{E_k}(x)$ and $g(t,x) = \sum_k b_k(t)\chi_{E_k}(x)$ where E_k are pairwise disjoint measurable sets, a_k are complex numbers and $b_k(t)$ are $\mathbb{C}^n$-valued, and set $B_k = \left(\int_0^\infty |b_k(t)|^2 dt\right)^{1/2}$.

Let

$$f_z(x) = \sum_k |a_k|^{\alpha(z)} \frac{a_k}{|a_k|} \chi_{E_k}(x)$$

---

[4] In practice, quadratic functionals are built from operators whose individual boundedness properties are known. The converse, thus, is never considered, so that this result appears new. Note that the argument could be written for arbitrary quadratic functionals made after an analytic family of operators. We leave to the reader the care of stating this general result.

and
$$g_z(t,x) = \sum_k B_k{}^{\beta(z)} \frac{b_k(t)}{B_k} \chi_{E_k}(x)$$
with
$$\alpha(z) = \frac{q}{p}(1-z) + \frac{q}{2}z \quad \text{and} \quad \beta(z) = \frac{q'}{p'}(1-z) + \frac{q'}{2}z.$$

Pick $\beta \in (0, \frac{\pi}{2} - \omega)$ and consider the function
$$F(z) = \int_{\mathbb{R}^n} \int_0^\infty \nabla e^{-te^{i\beta z}L} f_z(x) \cdot g_z(t,x)\,dxdt.$$
defined for z in the strip $0 \leq \Re z \leq 1$. This function is clearly continuous, and it is analytic in the interior of this strip. Moreover, for $z = iy$ with $y \in \mathbb{R}$, one finds easily from Hölder inequalities and the change of variable $te^{-\beta y} \mapsto t$,
$$|F(iy)| \leq \|G_{e^{-\beta y}L}(f_{iy})\|_p \|g_{iy}\|_{p'} \leq C_0 e^{\beta y/2} \|f_{iy}\|_p \|g_{iy}\|_{p'} = C_0 e^{\beta y/2}$$
and for $z = 1 + iy$ with $y \in \mathbb{R}$, by the square function estimate (6.2) for $e^{i\beta}L$ and the same change of variable,
$$|F(1+iy)| \leq \|G_{e^{i\beta(1+iy)}L} f_{1+iy}\|_2 \|g_{1+iy}\|_2 \leq C_1 e^{\beta y/2} \|f_{1+iy}\|_2 \|g_{1+iy}\|_2 = C_1 e^{\beta y/2}.$$
Here, C_1 depends on β in a non decreasing manner. Hence, the extension of three lines theorem in [64, Chapter V, Lemma 4.2] yields for $0 < x < 1$ and $y \in \mathbb{R}$,
$$|F(x+iy)| \leq C_0^{1-x} C_1^x e^{\beta y/2}.$$
This implies that L^q boundedness of $G_{e^{i\beta c_q}L}$ when $\Re c_q = \frac{\frac{1}{q} - \frac{1}{p}}{\frac{1}{2} - \frac{1}{p}}$ with a bound controlled by $Ce^{\beta \Im c_q/2}$. Now, this is true for all β in $(0, \frac{\pi}{2} - \omega)$ and also in $(-\frac{\pi}{2} + \omega, 0)$ by changing β to $-\beta$.

Keep q as before and let f, g be simple functions as above. If β is small enough, the function
$$G(z) = \int_{\mathbb{R}^n} \int_0^\infty \nabla e^{-te^{i\beta z}L} f(x) \cdot g(t,x)\,dxdt$$
is continuous in the strip $-1 \leq \Re z \leq 1$ and, by the previous argument, analytic in the open strip minus the real axis. Thus it is analytic in the open strip by Morera's theorem. Applying the three lines theorem again (this times, f and g are fixed and the exponent remains q) or Cauchy's theorem about 0 (since we are not after optimal bounds), we deduce a bound for $G'(0)$, which is equivalent to the L^q boundedness of $\widetilde{G}_L$.

Proof of Step 8: Reverse L^p inequality for G_L when $1 < p < +\infty$. What we have proved so far applies to any operator L in our class, and in particular, to $L = -\Delta$. The explicit formula for the heat kernel implies that $p_-(-\Delta) = 1$ and $q_+(-\Delta) = \infty$. Hence, we recover from our method the well-known estimate [5]
$$\|G_{-\Delta}(f)\|_p \lesssim \|f\|_p$$
for all $1 < p < \infty$.

---

[5] due to Stein. See [62].

Now, let $f, g \in L^2$ and observe that [6]

$$\int_{\mathbb{R}^n} f\,\overline{g} = \lim_{\varepsilon \downarrow 0} \int_{\mathbb{R}^n} e^{-\varepsilon L} f \, \overline{e^{\varepsilon \Delta} g} - \lim_{R \uparrow \infty} \int_{\mathbb{R}^n} e^{-RL} f \, \overline{e^{R\Delta} g}$$

$$= -\int_0^\infty \frac{d}{dt} \int_{\mathbb{R}^n} (e^{-tL}f)(x) \, \overline{(e^{t\Delta}g)(x)} \, dx \, dt$$

$$= \iint_{\mathbb{R}^n \times (0,\infty)} (A(x) + I)(\nabla e^{-tL}f)(x) \cdot \overline{(\nabla e^{t\Delta}g)(x)} \, dx dt.$$

The last equality is obtained by integration by parts in the x variable after computing the time derivative. Hence, we obtain

$$\left| \int_{\mathbb{R}^n} f\,\overline{g} \right| \leq (\|A\|_\infty + 1) \int_{\mathbb{R}^n} G_L(f) G_{-\Delta}(g).$$

Thus, if $1 < p < \infty$, the $L^{p'}$ boundedness of $G_{-\Delta}$ yields

$$\left| \int_{\mathbb{R}^n} f\,\overline{g} \right| \lesssim \|G_L(f)\|_p \|g\|_{p'}$$

and it follows

$$\|f\|_p \lesssim \|G_L(f)\|_p.$$

Of course this inequality is meaningfull whenever the right hand side is finite. □

Let us draw some consequences of our results.

COROLLARY 6.3. *If $n = 1$, we have for $1 < p < \infty$,*

$$\|G_L(f)\|_p \sim \|f\|_p \sim \|g_L(f)\|_p.$$

COROLLARY 6.4. *If $n = 2$, we have $\|g_L(f)\|_p \sim \|f\|_p$ for $1 < p < \infty$ and $\|G_L(f)\|_p \sim \|f\|_p$ for $1 < p < 2 + \varepsilon'$.*

COROLLARY 6.5. *If $n \geq 3$, we have $\|g_L(f)\|_p \sim \|f\|_p$ for $\frac{2n}{n+2} - \varepsilon < p < \frac{2n}{n-2} + \varepsilon'$ and $\|G_L(f)\|_p \sim \|f\|_p$ for $\frac{2n}{n+2} - \varepsilon < p < 2 + \varepsilon_1'$.*

COROLLARY 6.6. *If $n \geq 3$ and L has **real** coefficients, then we have $\|g_L(f)\|_p \sim \|f\|_p$ for $1 < p < \infty$ and $\|G_L(f)\|_p \sim \|f\|_p$ for $1 < p < 2 + \varepsilon$.*

COROLLARY 6.7. *If $n \geq 5$, there exists L such that g_L and G_L are not bounded for some p close to 1 (hence g_{L^*} is not bounded on L^p for some p close to ∞).*

We finish this section by proving Lemma 4.14.

PROOF OF LEMMA 4.14. The assumption of the lemma is that $(e^{-tL})_{t>0}$ satisfies $L^\rho - L^2$ off-diagonal estimates. By duality, $(e^{-tL^*})_{t>0}$ satisfies $L^2 - L^{\rho'}$ off-diagonal estimates. Hence, it follows from the method of step 2 above (applied to discrete times 4^k, $k \in \mathbb{Z}$, and changing $(tL)^{1/2} e^{-tL}$ to $\psi(tL)$ with ψ given by (4.15)) that for $2 < q' < \rho'$,

$$\left\| \left(\sum_{k \in \mathbb{Z}} |\psi(4^k L^*) f|^2 \right)^{1/2} \right\|_{q'} \leq C \|f\|_{q'}.$$

By duality, we obtain (4.2). □

---

[6] we learned this trick from J. Pipher in an unpublished manuscript. See [13] where it is used.

6.2. On inequalities of Stein and Fefferman for non-tangential square functions

Consider the functional

$$g_\lambda^*(f)(x) = \left(\iint_{\mathbb{R}_+^{n+1}} \left(\frac{t}{|x-y|+t} \right)^{n\lambda} |t\nabla u(y,t)|^2 \frac{dy\,dt}{t^{n+1}} \right)^{1/2}$$

where $u(x,t)$ is the harmonic extension of f and $1 < \lambda$. It is bounded on L^p ($1 < p < \infty$) if and only if $\lambda > \frac{2}{p}$.[7] At the critical case $\lambda = \frac{2}{p}$, it is weak-type (p,p).[8] Of course, the main tool for this is Calderón-Zygmund decomposition for L^p functions.

We show how to use our technology in this situation by replacing functions of the Laplacian by functions of L. Again, the main point is that L^p boundedness of the semigroup suffices. We also separate the cases $p < 2$ and $p > 2$.

Define

$$g_\lambda^*(f)(x)^2 = \iint_{\mathbb{R}_+^{n+1}} \left(\frac{\sqrt{t}}{|x-y|+\sqrt{t}} \right)^{n\lambda} |\mathcal{T}(e^{-tL}f)(y)|^2 \frac{dy\,dt}{\sqrt{t}^n},$$

where for fixed $t > 0$,

$$|\mathcal{T}g(y,t)|^2 = |\nabla_y g(y,t)|^2 + |L^{1/2}g(y,t)|^2.$$

In particular, this square function controls non tangential square functions where integration is performed on parabolic cones $|x-y| \le c\sqrt{t}$:

$$g_{NT}(f)(x)^2 = \iint_{|x-y|<c\sqrt{t}} |\mathcal{T}e^{-tL}f(y)|^2 \frac{dy\,dt}{\sqrt{t}^n}.$$

PROPOSITION 6.8. *If $q_-(L) < p < 2$ and $\lambda = \frac{2}{p}$ then g_λ^* has weak type (p,p).*

PROPOSITION 6.9. *If $2 < p < q_+(L)$ and $\lambda > 1$ then g_λ^* has strong type (p,p).*

COROLLARY 6.10. *If $q_-(L) < p < q_+(L)$ and $\lambda > \frac{2}{p}$, then g_{NT}, g_λ^* are bounded on L^p and one has* [9]

$$\|g_\lambda^* f\|_p \sim \|g_{NT}(f)\|_p \sim \|f\|_p.$$

Note that the result for g_{NT} holds for $q_-(L) < p < q_+(L)$ since it is independent of λ. The proof of the corollary is simple: we have a pointwise control of g_{NT} by g_λ^* for any $\lambda > \frac{2}{p}$ and we obtain boundedness. The reverse inequalities are obtained as for g_L and G_L following the heuristic idea below. Suppose we can write $\int f(y)g(y)\,dy = \iint f_t(y)g_t(y)\,dy\,dt$ and that $\int h(x,y,t)\,dx = 1$ for all y,t. Then

$$\int f(y)g(y)\,dy = \int \left(\iint f_t(y)g_t(y)h(x,y,t)\,dy\,dt \right) dx.$$

It suffices to apply Cauchy-Schwarz in the variables y,t to obtain non-tangential square function by choosing h and then Hölder in the x variable to deduce reverse inequalities from direct ones. We skip further details.

---

[7] This is due to Stein [62].

[8] This is in C. Fefferman's thesis [36].

[9] While Le Merdy's theorem mentioned above works well for the vertical square function of abstract operators on L^p spaces, it is not clear it applies to non-tangential square functions which are more geometrical objects in an abstract setting.

6.2. NON-TANGENTIAL SQUARE FUNCTIONS

REMARK. The limitations on p are only due to the presence of the spatial gradient ∇_y in the definition of g_λ^*. If one drops this gradient to keep only the $L^{1/2}$ part then the range of p becomes $p_-(L) < p < p_+(L)$.

PROOF OF PROPOSITION 6.8. Due to the fact that x and y may be far apart, Theorem 1.1 does not apply directly and one has to do some transformations. The key identity of this proof is that for any closed set F,

$$\int_F g_\lambda^*(f)(x)^2\, dx = \iint_{\mathbb{R}_+^{n+1}} |\mathcal{T}(e^{-tL}f)(y)|^2 J_{\lambda,F}(y,t)\, dydt$$

with

$$J_{\lambda,F}(y,t) = \frac{1}{\sqrt{t}^n} \int_F \left(\frac{\sqrt{t}}{|x-y|+\sqrt{t}}\right)^{n\lambda} dx.$$

First $J_{\lambda,F} \leq C$ so that

$$\|g_\lambda^*(f)\|_2^2 \leq C\|g_L(f)\|_2^2 + C\|G_L(f)\|_2^2$$

where g_L and G_L are the square functions defined earlier. The L^2 boundedness of g_L and G_L implies the L^2 boundedness of g_λ^*.

Second, we have also if y lies in some cube Q and $F = \mathbb{R}^n \setminus (2Q)$ that

$$J_{\lambda,F}(y,t) \leq C\sqrt{t}^{n(\lambda-1)} |Q|^{-(\lambda-1)/2}.$$

We begin as in Theorem 1.1, by looking at $\{g_\lambda^*(f) > \alpha\}$ and decomposing $f = g + \sum b_i$ according to the threshold α^p for $|f|^p$. For g use the L^2 boundedness of g_λ^*.

Next write again

$$b_i = A_{r_i} b_i + (1 - A_{r_i}) b_i$$

where A_r is the operator that works for g_L and G_L in the previous section. The term $\sum A_{r_i} b_i$ is again in L^2 with the right bound so that the L^2 boundedness of g_λ^* suffices again. It remains to estimate the size of the set

$$\{x \in \mathbb{R}^n; g_\lambda^*(\sum (1 - A_{r_i}) b_i)(x) > \alpha/3\}.$$

Again we take away the union of the dilated Whitney cubes $4Q_i$, whose mass is under control. It remains to estimate what is left on its complement F. By Tchebytchev's inequality, it is enough to estimate $\int_F g_\lambda^*(\sum (1-A_{r_i})b_i)(x)^2\, dx$ which we rewrite as

$$\iint_{\mathbb{R}_+^{n+1}} |\sum_i \mathcal{T}(e^{-tL}(1-A_{r_i})b_i)(y)|^2 J_{\lambda,F}(y,t)\, dydt.$$

The non local part of the ith summand is when $y \notin 2Q_i$. We bound $J_{\lambda,F}$ by a constant and we are back to the estimates performed to obtain the weak type (p,p) for g_L and G_L. We refer the reader to steps 3 and 4 in the previous section.

It remains to localise each ith summand on $2Q_i$. By the bounded overlaps of $2Q_i$'s [10] and the second upper bound on $J_{\lambda,F}$, we have an upper bound

$$N \sum_i \iint_{2Q_i \times \mathbb{R}^+} \frac{\sqrt{t}^{n(\lambda-1)}}{|Q_i|^{(\lambda-1)/2}} |\mathcal{T}(e^{-tL}(1-A_{r_i})b_i)(y)|^2\, dydt.$$

---

[10] We have to make sure in the construction of the Whitney cubes that this actually holds and may be twice the cubes is not appropriate but certainly cQ_i with some $c > 1$ is.

Now for each i, we integrate on the full upper half space: the integral in y (with t fixed) and the solution of the Kato problem [11] allow us to bound the term with ∇_y by the one with $L^{1/2}$. Next if $a = \frac{n}{2}(\lambda - 1) = \frac{n}{p} - \frac{n}{2}$, the L^p bounds of the semigroup imply by Proposition 5.3 the Hardy-Littlewood-Sobolev inequality

$$\|L^{-a} f\|_2 \leq C \|f\|_p.$$

Using the square function estimate of McIntosh-Yagi based on $(tL)^a e^{-tL}$ we obtain that the ith term is bounded by

$$C|Q_i|^{1-2/p} \|L^{-a}(1 - A_{r_i}) b_i\|_2^2 \leq C|Q_i|^{1-2/p} \|(1 - A_{r_i}) b_i\|_p^2 \leq C|Q_i|^{1-2/p} \|b_i\|_p^2.$$

It remains to sum other i and we are done. $\square$

PROOF OF PROPOSITION 6.9. Here also, we cannot apply directly Theorem 1.2 but rather its spirit and its proof. We let $f \in L^2(\mathbb{R}^n)$ and B be a ball with radius r. We also let $2 < p_0 < q_+(L)$ and $m > n/4$ an integer. Assume that we have proved that for $k = 1, 2, \ldots, m$

$$(6.5) \qquad \left(\frac{1}{|B|} \int_B |g_\lambda^*(e^{-kr^2 L} f)|^{p_0}\right)^{1/p_0} \leq C \inf_{x \in B} M(g_\lambda^*(f)^2)^{1/2}(x)$$

then we can argue as follows. For $A_r = I - (I - e^{r^2 L})^m$, we have

$$g_\lambda^*(f)^2(x) \leq 2 g_\lambda^*(A_r f)^2(x) + 2 g_\lambda^*((I - A_r) f)^2(x).$$

Write

$$g_\lambda^*((I - A_r) f)^2(x) = \iint_{y \in 2B} h(x - y, t) |\mathcal{T}(e^{-tL}(I - A_r) f)(y)|^2 dy dt$$
$$+ \iint_{y \notin 2B} h(x - y, t) |\mathcal{T}(e^{-tL}(I - A_r) f)(y)|^2 dy dt$$

with $h(x, t) = \sqrt{t}^n \left(\frac{\sqrt{t}}{|x| + \sqrt{t}}\right)^{n\lambda}$. As $h(x - y, t) \sim h(z - y, t)$ for $x, z \in B$ and $y \notin 2B$, the second integral is bounded by

$$2 g_\lambda^*(A_r f)^2(x) + 2 \inf_{z \in B} g_\lambda^*(f)^2(z).$$

Hence, we can apply Proposition 1.5 as in the proof of Theorem 1.2 with

$$G_B(x) = 2 \iint_{y \in 2B} h(x - y, t) |\mathcal{T}(e^{-tL}(I - A_r) f)(y)|^2 dy dt$$

and

$$H_B(x) = 4 g_\lambda^*(A_r f)^2(x) + 2 \inf_{z \in B} g_\lambda^*(f)^2(z)$$

provided we show that

$$\frac{1}{|B|} \int_B G_B \leq C \inf_{z \in B} M(|f|^2)(z).$$

But, $\int_B h(x - y, t) \, dx \leq 1$ since $\lambda > 1$, hence

$$\int_B G_B(x) \, dx \leq 2 \iint_{2B \times \mathbb{R}^+} |\mathcal{T}(e^{-tL}(I - A_r) f)(y)|^2 dy dt$$

---

[11] One can proceed also using Remark 5.2.

and we are back to the calculations made in steps 1 and 2 of the previous section that give us
$$\iint_{2B\times\mathbb{R}^+} |\mathcal{T}(e^{-tL}(I-A_r)f)(y)|^2 dy dt \leq C \inf_{z\in B} M(|f|^2)(z).$$

Hence, it remains to establish (6.5). To do that, assume that B is the unit ball and $r=1$ (One can treat the general case by rescaling and translation, which changes L to another operator with the same properties and the same critical numbers). Assume also for simplicity that $k=1$. We write for $x\in B$,
$$g^*_\lambda(e^{-r^2L}f)^2(x) = I + II$$
where
$$I = \iint_{(y,t)\notin E} h(x-y,t)|\mathcal{T}(e^{-tL}e^{-r^2L}f)(y)|^2 dy dt$$
and
$$II = \iint_{(y,t)\in E} h(x-y,t)|\mathcal{T}(e^{-tL}e^{-r^2L}f)(y)|^2 dy dt$$
and $(y,t)\notin E$ means $y\in 2B$ and $t\leq r(2B)^2$. Let us treat the first term.

Using Minkowksi integral inequality (with respect to t) and $\int_{2B} h(x-y,t)\,dx \leq 1$ whenever $y\in 2B$, we obtain
$$\left(\int_B I^{p_0/2}\right)^{2/p_0} \leq \int_0^4 \left(\int_{2B} |\mathcal{T}(e^{-r^2L}e^{-tL}f)(y)|^{p_0} dy\right)^{2/p_0} dt.$$
Following again the calculations in steps 1 and 2 of the previous section
$$\left(\int_{2B} |\mathcal{T}(e^{-tL}(I-A_r)f)(y)|^{p_0} dy\right)^{2/p_0} \leq \sum_{j\geq 2} \frac{c_j}{|2^{j+1}B|} \int_{2^{j+1}B} |\mathcal{T}(e^{-tL}f)(y)|^2 dy$$
with $c_j = Ce^{-c4^j}$ and with the limitation $p_0 < q_+(L)$ from ∇_y and $p_0 < p_+(L)$ from $L^{1/2}$ in the definition of $\mathcal{T}$. Now since $t\leq 4$ and $j\geq 2$,
$$\int_{2^{j+1}B} |\mathcal{T}(e^{-tL}f)(y)|^2 dy dt \leq \frac{A}{\sqrt{t}^n} \int_{2^{j+1}B} \int_{|x-y|\leq \sqrt{t}} |\mathcal{T}(e^{-tL}f)(y)|^2 dy\, dx$$
for some $A>0$. Hence, we obtain readily
$$\left(\int_B I^{p_0/2}\right)^{2/p_0} \leq AC \inf_{x\in B} M((g_{NT}f)^2)^{1/2}(x).$$

For the second term II, we first observe that for $x\in B$ and $(y,t)\in E$, then $h(x-y,t) \sim h(y,t)$ (recall that B is the unit ball so that 0 is its center), thus $\left(\int_B II^{p_0/2}\right)^{2/p_0} \leq \sup_B II$. Next, decompose E as the union of E_k, $k\geq -1$, as follows: $E_{-1} = E \cap \{|y|<\sqrt{t}\}$ and $E_k = E\cap\{2^k\sqrt{t}<|y|\leq 2^{k+1}\sqrt{t}\}$. Then for $(y,t)\in E_k$, $h(y,t) \leq 2^{-kn\lambda}\sqrt{t}^{-n}$. A crucial geometrical observation is that if $(y,t)\in E_k$ then $2^k\sqrt{t}\geq 1$. Using the method of the second argument in the proof of Theorem 4.6 (recall that $r=1$),
$$\int_{|y|\leq 2^k\sqrt{t}} |\mathcal{T}e^{-r^2L}(e^{-tL}f)(y)|^2 dy \leq C\int_{|z|\leq 2^{k+1}\sqrt{t}} |\mathcal{T}(e^{-tL}f)(z)|^2 dz$$
$$+ \sum_{j\geq k+2} \int_{|z|\leq 2^j\sqrt{t}} Ce^{-c4^j t}|\mathcal{T}(e^{-tL}f)(z)|^2 dz.$$

Hence
$$II \leq \iint m(z,t)|T(e^{-tL}f)(z)|^2 dzdt$$
where
$$m(z,t) = C\sum_{k\geq -1} 2^{-kn\lambda}\sqrt{t}^{-n}1_{2^k\sqrt{t}\geq 1}(1_{|z|\leq 2^{k+1}\sqrt{t}} + \sum_{j\geq k+2} e^{-c4^j t}1_{|z|\leq 2^j\sqrt{t}}).$$

Tedious but elementary verifications show that $m(z,t) \leq Ch(x-z,t)$ for all $x \in B$ and $(z,t) \in \mathbb{R}^{n+1}_+$ using only $\lambda > 0$. Hence, $II \leq C\inf_{x\in B} g^*_\lambda(f)^2(x)$. □

CHAPTER 7

Miscellani

7.1. Local theory

Let L be as in the Introduction. We have developed a global (or homogeneous) L^p theory by making global in time assumptions on the semigroup. Reasons for this theory not to apply to a particular L at p are that the semigroup is not L^p bounded for some (or all) $t > 0$ (in which case this is the end of the story) or that the semigroup operators are bounded on L^p but not uniformly, ususally with an exponential blow up. In the second case, adding a large s to L gives us back the uniformity. The local L^p theory consists in working with $L + s$ instead of L. Hence, the above results may be adapted with minor modifications in the proofs by changing systematically L to $L + s$ for $s > 0$. One may define the four critical exponents $p_\pm(L+s)$ and $q_\pm(L+s)$ which may depend on s or not. Indeed, the L^2 theory developed in Chapter 2 works with $s = 0$ and the numbers $p_-(L+s)$, $q_-(L+s)$ are non increasing, and $p_+(L+s)$, $q_+(L+s)$ non decreasing as s grows.

We have the following assertions for $s > 0$.

(1) $p_-(L+s) = q_-(L+s)$ and $(q_+(L+s))^* \leq p_+(L+s)$.
(2) $q_+(L+s)$ is the supremum of exponents p for which one has the invertibility of $L+s$ from $W^{1,p}$ onto $W^{-1,p}$.
(3) For the Riesz transform $\nabla(L+s)^{-1/2}$ the range for L^p boundedness is the open interval $(p_-(L+s), q_+(L+s))$. For p in this range $L+s$ is an isomorphism from $W^{1,p}$ onto L^p.
(4) There is bounded holomorphic functional calculus for $L+s$ on L^p essentially when $p_-(L+s) < p < p_+(L+s)$.
(5) There is a Hardy-Littlewood-Sobolev-Kato diagram.
(6) There is an equivalent L^p norm defined by g_{L+s} essentially for $p_-(L+s) < p < p_+(L+s)$.
(7) There is an equivalent L^p norm defined by G_{L+s} essentially for $q_-(L+s)' < p < q_+(L+s)$.

This applies to operators whose coefficients have some smoothness. If the coefficients are, in addition, BUC (bounded uniformly continuous) or in the closure of BUC for the bmo norm, then it is known that $p_-(L+s) = q_-(L+s) = 1$ and $p_+(L+s) = q_+(L+s) = \infty$ for s large enough.[1] This gives L^p estimates for Riesz transforms, functional calculi, square functions in the range $1 < p < \infty$.

One can also add to L perturbation by lower order terms and develop the similar theory.

---

[1] This is a consequence of [2] (see also [44])

One can probably develop this theory for operators on domains with Lipschitz boundaries at least with Dirichlet or Neumann boundary conditions. This is left to the interested reader.

Another interesting direction is to test this theory for other classes of elliptic operators such as Schrödinger operators for which criteria for the determination of $p_-(L)$ and $p_+(L)$ have been given. [2] This theory already applies in the range $p < 2$. [3] It remains to study the range $p > 2$.

7.2. Higher order operators and systems

Consider an homogeneous elliptic operator L of order m, $m \in \mathbb{N}$, $m \geq 2$, defined by

$$(7.1) \qquad Lf = (-1)^m \sum_{|\alpha|=|\beta|=m} \partial^\alpha(a_{\alpha\beta} \partial^\beta f),$$

where the coefficients $a_{\alpha\beta}$ are complex-valued L^∞ functions on $\mathbb{R}^n$, and we assume

$$(7.2) \qquad \left| \sum_{|\alpha|=|\beta|=m} \int_{\mathbb{R}^n} a_{\alpha\beta}(x) \partial^\beta f(x) \partial^\alpha \bar{g}(x)\, dx \right| \leq \Lambda \|\nabla^m f\|_2 \|\nabla^m g\|_2$$

and the strong Gårding inequality

$$(7.3) \qquad \mathrm{Re} \sum_{|\alpha|=|\beta|=m} \int_{\mathbb{R}^n} a_{\alpha\beta}(x) \partial^\beta f(x) \partial^\alpha \bar{f}(x)\, dx \geq \lambda \|\nabla^m f\|_2^2$$

for some $\lambda > 0$ and $\Lambda < +\infty$ independent of $f, g \in W^{m,2}$. Here, ∇^k is the array of all kth order derivatives.

One can also generalize second order or higher order operators to elliptic systems of any even order verifying the strong Gårding inequality. For simplicity of exposition we stick to the scalar case but all works similarly for systems.

The L^2 theory for the semigroup is analogous. There are
  (1) bounded holomorphic functional calculus on L^2,
  (2) L^2 off-diagonal estimates for the families $(t^{k/2m} \nabla^k e^{-tL})_{t>0}$ and their analytic extensions for $0 \leq k \leq m$ where the gaussian decay e^{-cu^2} is changed to $e^{-cu^{\frac{2m}{2m-1}}}$ and the homogeneity changes from $\sqrt{t}$ to $t^{1/2m}$.

These estimates yield the generalized conservation property

$$e^{-tL} P = P$$

for all $t > 0$ in the L^2_{loc} sense for P polynomial of degree less than m. [4] Since L is constructed as before as a maximal-accretive operator, it has a square root and one has in all dimensions [5]

$$(7.4) \qquad \|L^{1/2} f\|_2 \sim \|\nabla^m f\|_2.$$

---

[2] See [50]

[3] [20] and [21] for results on the functional calculus and Riesz transforms

[4] See [11] for a proof under additional hypotheses. The argument has the same structure in the general case.

[5] This is [11], Theorem 1.1.

Moreover, the square functions g_L and G_L define equivalent norms on L^2 (in G_L replace ∇ by ∇^m).

Then one can develop the L^p theory of the semigroup, introducing the limits $p_\pm(L)$ for the L^p boundedness of $(e^{-tL})_{t>0}$ and the limits $q_\pm(L)$ for the L^p boundedness of $(\sqrt{t}\,\nabla^m e^{-tL})_{t>0}$. The results are similar with more technical burden in the arguments as one often has to control intermediate families $(t^{k/2m}\nabla^k e^{-tL})_{t>0}$ for $1 \le k \le m-1$.[6] One has that

$$p_-(L) = q_-(L)$$
$$p_+(L) \ge (q_+(L))^{*m}$$

where p^{*m} means m times the operation $p \mapsto p^*$.

By Sobolev embeddings plus perturbation results (such as Lemma 3.4) we have

$$p_-(L) \begin{cases} = 1, & \text{if } n \le 2m \\ < \frac{2n}{n+2m}, & \text{if } n > 2m. \end{cases}$$

Furthermore, this upper bound is sharp for the class of **all** higher order operators with $n > 2m \ge 4$: for any n and $m \ge 2$ with $2m < n$ and $p < \frac{2n}{n+2m}$, there exists an operator (scalar or system) L of order $2m$ such that $p_-(L) > p$.[7] One has also

$$q_+(L) \begin{cases} = +\infty, & \text{if } n = 1 \\ > 2, & \text{if } n \ge 2. \end{cases}$$

If L^* is an example with $p_-(L^*) \sim \frac{2n}{n+2m}$ ($n > 2m$), then $p_+(L) \sim \frac{2n}{n-2m}$ and one has $p_+(L) \sim (q_+(L))^{*m}$ (Here, $\sim$ means within some arbitrary small ε). Thus the inequality $p_+(L) \ge (q_+(L))^{*m}$ is best possible.

Next, the Riesz transform becomes $\nabla^m L^{-1/2}$ and

$$\|\nabla^m L^{-1/2} f\|_p \lesssim \|f\|_p \quad \text{if and only if} \quad q_-(L) < p < q_+(L)$$

and one can show the reverse inequalities (for some $0 < c < \pi/2$ for all $t \in \mathbb{R}$)

$$\|L^{1/2+it} f\|_p \lesssim e^{c|t|} \|\nabla^m f\|_p \quad \text{whenever} \quad \sup(1, (p_-(L))_{*m}) < p < p_+(L).$$

The bounds for $p > 2$ are merely obtained by duality from the Riesz transform bounds and $p_+(L)$ is best possible. A tool to obtain the estimates for $p < 2$ is the extension of the Calderón-Zygmund decomposition to Sobolev $W^{m,p}$ functions.[8] The lower limit $(p_-(L))_{*m}$ (if not ≤ 1, which implies large dimensions), is best possible if L is an operator for which $p_+(L^*) \ge q_+(L^*)^{*m}$ is best possible. We have seen there exist such operators. The connexion with Hodge theory is analogous to the second order case and the dichotomy $p > 2$ vs $p < 2$ appears again.

The critical numbers $\frac{2n}{n\pm 2}$ and $\frac{2n}{n+4}$ which appear in the L^p theory of square roots for second order operators (namely Propositions 4.18 and 4.19) become $\frac{2n}{n\pm 2m}$ and $\frac{2n}{n+4m}$.[9] Hence, the range of exponents p for the Riesz transform L^p estimate

---

[6] see [**17**], Chapter I, [**16**] and [**29**]

[7] This is in Davies [**30**] based on examples of Maz'ya and de Giorgi.

[8] See [**6**] for the proof.

[9] See [**6**] where all this is explicited. The case $\frac{2n}{n+2m} < p < 2$ when $n > 2m$ is due to Blunck & Kunstmann [**20**]. The other cases follow from the methods in [**17**] although this is not explicited.

is
$$\begin{cases} 1 < p < \infty, & \text{when } n = 1, m \geq 1, \\ 1 < p < 2 + \varepsilon, & \text{when } 1 < n \leq 2m, \\ \frac{2n}{n+2m} - \varepsilon < p < 2 + \varepsilon', & \text{when } n > 2m. \end{cases}$$

The discussion above show that these open ranges are best possible. The range of exponents p for the reverse inequality [10] is

$$\begin{cases} 1 < p < \infty, & \text{when } n \leq 2m, \\ 1 < p < \frac{2n}{n-2m} + \varepsilon, & \text{when } 2m < n \leq 4m, \\ \frac{2n}{n+4m} - \varepsilon_1 < p < \frac{2n}{n-2m} + \varepsilon, & \text{when } n > 4m. \end{cases}$$

Again, these open ranges are best possible. [11]

The bounded holomorphic functional calculus extends on L^p for $p_-(L) < p < p_+(L)$. [12]

The theory of square functions also generalize similarly. [13] For g_L the range of p's is $p_-(L) < p < p_+(L)$ and for G_L, $q_-(L) < p < q_+(L)$. One could treat also variants of the non-tangential square functions.

If the strong Gårding inequality is weakened by a term $-\kappa \|f\|_2^2$ in the right hand side (in both the operator and system cases), then one has to replace $L + \lambda$ for $\lambda \geq \kappa$ and the local theory applies.

One can also add pertubation by lower order terms with bounded measurable coefficients without any harm to the theory.

---

[10] This is proved in [6] for $p < 2$ and $n > 4m$. The other cases were done earlier as a consequence of the methods in [17] for $n \leq 2m$ and in [10] for $2m \leq n < 4m$.

[11] In particular, this means that the methods used here cannot improve the ranges of p for second order operators (as improved methods would apply for higher order as well) unless they use specific features of second order operators.

[12] This is due to Blunck & Kunstmann [21].

[13] For g_L, as for second order operators, one can also combine results of Blunck & Kunstmann and of Le Merdy; for G_L this is new.

APPENDIX A

Calderón-Zygmund decomposition for Sobolev functions

Here we prove the Calderón-Zygmund decomposition for Sobolev functions. The notation are those of the statement.

PROOF OF LEMMA 4.12. If $p = \infty$, set $g = f$. Assume next that $p < \infty$. Let $\Omega = \{x \in \mathbb{R}^n; M(|\nabla f|^p)(x) > \alpha^p\}$ where M is the uncentered maximal operator over cubes of $\mathbb{R}^n$. If Ω is empty, then set $g = f$. Otherwise, the maximal theorem gives us

$$|\Omega| \leq C\alpha^{-p} \int_{\mathbb{R}^n} |\nabla f|^p.$$

Let F be the complement of Ω. By the Lebesgue differentiation theorem, $|\nabla f| \leq \alpha$ almost everywhere on F. We also have,

LEMMA A.1. *One can redefine f on a null set of F so that for all $x \in F$, for all cube Q centered at x,*

(A.1) $$|f(x) - m_Q f| \leq C\alpha \ell(Q)$$

where $\ell(Q)$ is the sidelength of Q and for all $x, y \in F$,

(A.2) $$|f(x) - f(y)| \leq C\alpha |x - y|.$$

The constant C depends only on dimension and p.

Here $m_E f$ denotes the mean of f over E. It is well-defined if E is a cube as f is locally integrable. Let us postpone the proof of this lemma and continue the argument.

Let (Q_i) be a Whitney decomposition of Ω by dyadic cubes. Hence, Ω is the disjoint union of the Q_i's, the cubes $2Q_i$ are contained in Ω and have the bounded overlap property, but the cubes $4Q_i$ intersect F. As usual, λQ is the cube co-centered with Q with sidelength λ times that of Q. Hence (4.10) and (4.11) are satisfied by the cubes $2Q_i$. Let us now define the functions b_i. Let $(\mathcal{X}_i)$ be a partition of unity on Ω associated to the covering (Q_i) so that for each i, $\mathcal{X}_i$ is a C^1 function supported in $2Q_i$ with $\|\mathcal{X}_i\|_\infty + \ell_i \|\nabla \mathcal{X}_i\|_\infty \leq c(n)$, ℓ_i being the sidelength of Q_i. Pick a point $x_i \in 4Q_i \cap F$. Set

$$b_i = (f - f(x_i))\mathcal{X}_i.$$

It is clear that b_i is supported in $2Q_i$. Let us estimate $\int_{2Q_i} |\nabla b_i|^p$. Introduce $\widetilde{Q}_i$ the cube centered at x_i with sidelength $8\ell_i$. Then $2Q_i \subset \widetilde{Q}_i$. Set $c_i = m_{2Q_i} f$ and $\tilde{c}_i = m_{\widetilde{Q}_i} f$ and write

$$b_i = (f - c_i)\mathcal{X}_i + (c_i - \tilde{c}_i)\mathcal{X}_i + (\tilde{c}_i - f(x_i))\mathcal{X}_i.$$

By (A.1) and (4.11) for the cubes $2Q_i$, $|\tilde{c}_i - f(x_i)| \leq C\alpha\ell_i$, hence $\int_{2Q_i} |\tilde{c}_i - f(x_i)|^p |\nabla \mathcal{X}_i|^p \leq C\alpha^p |2Q_i|$. Next, using the L^p-Poincaré inequality and the fact that $\widetilde{Q}_i \cap F$ is not empty,

$$|c_i - \tilde{c}_i| \leq \frac{1}{|2Q_i|} \int_{\widetilde{Q}_i} |f - \tilde{c}_i| \leq C\ell_i \left(\frac{1}{|\widetilde{Q}_i|} \int_{\widetilde{Q}_i} |\nabla f|^p \right)^{1/p} \leq C\alpha\ell_i.$$

Hence, $\int_{2Q_i} |c_i - \tilde{c}_i|^p |\nabla \mathcal{X}_i|^p \leq C\alpha^p |2Q_i|$. Lastly, since $\nabla\big((f - c_i)\mathcal{X}_i\big) = \mathcal{X}_i \nabla f + (f - c_i)\nabla \mathcal{X}_i$, we have again by the L^p-Poincaré inequality and the fact that the average of $|\nabla f|^p$ on $2Q_i$ is controlled by $C\alpha^p$ that

$$\int_{2Q_i} |\nabla\big((f - c_i)\mathcal{X}_i\big)|^p \leq C\alpha^p |2Q_i|.$$

Thus (4.9) is proved.

Set $h(x) = \sum_i f(x_i)\nabla\mathcal{X}_i(x)$. Note that this sum is locally finite and $h(x) = 0$ for $x \in F$. Note also that $\sum_i \mathcal{X}_i(x)$ is 1 on Ω and 0 on F. Since it is also locally finite we have $\sum_i \nabla\mathcal{X}_i(x) = 0$ for $x \in \Omega$. We claim that $h(x) \leq C\alpha$. Indeed, fix $x \in \Omega$. Let Q_j be the Whitney cube containing x and let I_x be the set of indices i such that $x \in 2Q_i$. We know that $\sharp I_x \leq N$. Also for $i \in I_x$ we have that $C^{-1}\ell_i \leq \ell_j \leq C\ell_i$ and $|x_i - x_j| \leq C\ell_j$ where the constant C depends only on dimension (see [**62**]). We have

$$|h(x)| = \left| \sum_{i \in I_x} (f(x_i) - f(x_j))\nabla\mathcal{X}_i(x) \right| \leq C \sum_{i \in I_x} |f(x_i) - f(x_j)|\ell_i^{-1} \leq CN\alpha,$$

by the previous observations.

It remains to obtain (4.7) and (4.8). We easily have using $\sum_i \nabla\mathcal{X}_i(x) = 0$ for $x \in \Omega$, that

$$\nabla f = (\nabla f)\mathbf{1}_F + h + \sum_i \nabla b_i, \quad \text{a.e.}.$$

Now $\sum_i b_i$ is a well-defined distribution on $\mathbb{R}^n$. Indeed, for a test function u, using the properties of the Whitney cubes,

$$\sum_i \int |b_i u| \leq C \int \left(\sum_i |b_i(x)|\ell_i^{-1} \right) |u(x)| d(x, F)\, dx$$

and the last sum converges in L^p as a consequence of (4.10) and

LEMMA A.2. *Set* $p^* = \frac{np}{n-p}$ *if* $p < n$ *and* $p^* = \infty$ *otherwise, then for all real numbers* r *with* $p \leq r \leq p^*$,

(A.3) $$\| \sum_i |b_i|\ell_i^{-1} \|_r^r \leq C\alpha^r \sum_i |Q_i|.$$

Admit this lemma and set $g = f - \sum_i b_i$. Then $\nabla g = (\nabla f)\mathbf{1}_F + h$ in the sense of distributions and, hence, ∇g is a bounded function with $\|\nabla g\|_\infty \leq C\alpha$. $\square$

PROOF OF LEMMA A.2. By (4.11) and the Poincaré-Sobolev inequality:

$$\| \sum_i |b_i|\ell_i^{-1} \|_r^r \leq N \sum_i \| |b_i|\ell_i^{-1} \|_r^r \leq NC \sum_i \ell_i^{r\theta} \|\nabla b_i\|_p^r$$

where $\theta = \frac{n}{r} - \frac{n}{p}$. By (4.9), $\ell_i^{r\theta}\|\nabla b_i\|_p^r \leq \alpha^r \ell_i^{nr/p}$, hence

$$\|\sum_i |b_i|\ell_i^{-1}\|_r^r \leq CN\alpha^r \sum_i \ell_i^n.$$

□

PROOF OF LEMMA A.1. Let x be a point in F. Fix such cube Q with center x and let Q_k be co-centered cubes with $\ell(Q_k) = 2^k\ell(Q)$ for k a negative integer. Then, by Poincaré's inequality

$$\begin{aligned}|m_{Q_{k+1}}f - m_{Q_k}f| &\leq 2^n |m_{Q_{k+1}}(f - m_{Q_{k+1}}f)| \\ &\leq C2^n \ell(Q_k)(m_{Q_{k+1}}|\nabla f|^p)^{1/p} \\ &\leq C2^k \ell(Q)\alpha\end{aligned}$$

since Q_{k+1} contains $x \in F$. It easily follows that $m_Q f$ has a limit as $|Q|$ tends to 0. If, moreover, x is in the Lebesgue set of f, then this limit is equal to $f(x)$. Redefine f on the complement of the Lebesgue set in F so that $m_Q f$ tends to $f(x)$ with Q centered at x with $|Q| \to 0$. Moreover, summing over k the previous inequality gives us (A.1). To see (A.2), let Q_x be the cube centered at x with sidelength $2|x-y|$ and Q_y be the cube centered at y with sidelength $8|x-y|$. It is easy to see that $Q_x \subset Q_y$. As before, one can see that $|m_{Q_x}f - m_{Q_y}f| \leq C\alpha|x-y|$. Hence by the triangle inequality and (A.1), one obtains (A.2) readily. □

Bibliography

[1] G. Alexopoulos. On the large time behavior of the heat kernels of quasi periodic differential operators, *J. of Geom. Anal.* **10** (2000), 207-218.

[2] J.M. Angeletti, S. Mazet & P. Tchamitchian. Analysis of second order elliptic operators without boundary conditions with VMO or Hölderian coefficients. In Dahmen, Kurdila, Oswald, editor, *Multiscale wavelets methods for partial differential equations*, volume 6 of *Wavelets and its Aplications*. Acad. Press, 1997.

[3] W. Arendt. Gaussian estimates and interpolation of the spectrum in L^p, *Diff. Int. Eqs* **7** (1994), 1153–1168.

[4] W. Arendt & A.F.M. ter Elst. Gaussian estimates for second order elliptic operators with boundary conditions, *J. Operator Theory* **38** no 1 (1997), 87–130.

[5] D. Aronson. Bounds for fundamental solutions of a parabolic equation. *Bull. Amer. Math. Soc.* **73** (1967), 890–896.

[6] P. Auscher. On L^p-estimates for square roots of second order elliptic operators on $\mathbb{R}^n$, *Publ. Mat.* **48** (2004), 159–186.

[7] P. Auscher, L. Barthélémy, P. Bénilan & El M. Ouhabaz. Absence de la L^∞ contractivité pour les semi-groupes associés aux opérateurs elliptiques complexes sous forme divergence. *Pot. Anal.* **12** no. 2 (2000), 169–189.

[8] P. Auscher, T. Coulhon, X.T. Duong & S. Hofmann. Riesz transforms on manifolds and heat kernel regularity, *Ann. Scient. ENS Paris* **37** no. 6 (2004), 911-957.

[9] P. Auscher, T. Coulhon & Ph. Tchamitchian. Absence de principe du maximum pour certaines équations paraboliques complexes, *Coll. Math.* **171** (1996), 87–95.

[10] P. Auscher, S. Hofmann, M. Lacey, A. McIntosh & Ph. Tchamitchian. The solution of the Kato square root problem for second order elliptic operators on $\mathbb{R}^n$, *Ann. Math.* (2) **156** (2002), 633–654.

[11] P. Auscher, S. Hofmann, A. McIntosh and Ph. Tchamitchian. The Kato square root problem for higher order elliptic operators and systems on $\mathbb{R}^n$, *J. Evol Equ.* **1** (2001), 361–385.

[12] P. Auscher & J.M. Martell, *Weighted norm inequalities, off-diagonal estimates and elliptic operators. Part II: Off-diagonal estimates*, Preprint 2006.

[13] P. Auscher, A. McIntosh & A. Nahmod. Holomorphic functional calculi of operators, quadratic estimates and interpolation, *Indiana Univ. Math. J.* **46** (1997), 375–403.

[14] P. Auscher, A. McIntosh & A. Nahmod. The square root problem of Kato in one dimension, and first order systems, *Indiana Univ. Math. J.* **46** (1997), 659–695.

[15] P. Auscher, A. McIntosh & Ph. Tchamitchian. Heat kernel of complex elliptic operators and applications, *J. Funct. Anal.* **152** (1998), 22–73.

[16] P. Auscher & M. Qafsaoui. Equivalence between regularity theorems and heat kernel estimates for higher order operators and systems under divergence form, *J. Funct. Anal.* **177** (2000), 310–364.

[17] P. Auscher & Ph. Tchamitchian. *Square root problem for divergence operators and related topics*, volume 249 of *Astérisque*, Soc. Math. France, 1998.

[18] P. Auscher & Ph. Tchamitchian. Calcul fonctionnel précisé pour des opérateurs elliptiques complexes en dimension un (et applications à certaines équations elliptiques complexes en dimension deux), *Ann. Inst. Fourier* **45** (1995), 721–778.

[19] A. Bensoussan, J.-.L. Lions, and G. Papanicolaou. *Asymptotic Analysis for periodic structures*. Studies in Math. & its Applications. North Holland, 1978.

[20] S. Blunck & P. Kunstmann. Calderón-Zygmund theory for non-integral operators and the H^∞-functional calculus, *Rev. Mat. Iberoamericana* **19** no. 3 (2003), 919–942.

[21] S. Blunck & P. Kunstmann. Weak-type (p,p) estimates for Riesz transforms, *Math. Z.* **247** no. 1 (2004), 137–148.

[22] L.A. Caffarelli & I. Peral. On $W^{1,p}$ estimates for elliptic equations in divergence form, *Comm. Pure App. Math.* **51** (1998), 1–21.

[23] A. P. Calderón. Commutators, singular integrals on Lipschitz curves and applications, *Proceedings of the I.C.M. Helsinki 1978* Acad. Sci. Fennica, Helsinki 1980, 85-96.

[24] A. P. Calderón & A. Zygmund. On the existence of certain singular integrals, *Acta Math.* **88** (1952), 85-139.

[25] R. Coifman, A. McIntosh & Y. Meyer. L'intégrale de Cauchy définit un opérateur borné sur $L^2(\mathbb{R})$ pour les courbes lipschitziennes. *Ann. Math.* **116** (1982), 361–387.

[26] R. Coifman & G. Weiss, Extensions of Hardy spaces and their use in analysis, *Bull. A. M. S.* **83** (1977) 569–645.

[27] T. Coulhon T. & X.T. Duong, Riesz transforms for $1 \leq p \leq 2$, *Trans. A. M. S.* **351** (1999), 1151-1169.

[28] M. Cowling, I. Doust, McIntosh & A. Yagi. Banach space operators with a bounded H^∞ functional calculus, *J. Aust. Math. Soc.*, (Series A), **60** (1996), 51–89.

[29] E.B. Davies. Uniformly elliptic operators with measurable coefficients, *J. Funct. Anal.* **132** (1995), 141–169.

[30] E.B. Davies. Limits on L^p regularity of selfadjoint elliptic operators, *J. Differential Equations* **135**, no. 1 (1997), 83–102.

[31] E.B. Davies. L^p spectral independence and L^1 analyticity, *J. London Math. Soc.* (2) **52** no. 1 (1995), 177–184.

[32] N. Dungey, A.F.M. ter Elst & D. Robinson. On second-order almost-periodic elliptic operators, *J. London Math. Soc.* (2) **63** (2001), 735–753.

[33] X.T. Duong & A. McIntosh. Singular integral operators with non-smooth kernels on irregular domains, *Rev. Mat. Iberoamericana*, **15** no. 2 (1999), 233–265.

[34] X.T. Duong & A. McIntosh. The L^p boundedness of Riesz transforms associated with divergence forms operators, Workshop on Analysis and Applications, Brisbane, 1997, Volume 37 of *Proceedings of the Centre for Mathematical Analysis*, ANU, Canberra, pp 15–25, 1999.

[35] A.F.M. ter Elst, D. Robinson & A. Sikora. On second-order periodic elliptic operators in divergence form, *Math. Z.* **238** (2001), 569–637.

[36] C. Fefferman. Inequalities for strongly singular convolution operators, *Acta Math.* **124** (1970), 9–36.

[37] C. Fefferman & E.M. Stein. H^p spaces in several variables, *Acta Math.* **129** (1972) 137–193.

[38] M. Giaquinta & G. Modica. Regularity results for some classes of higher order non linear elliptic systems, *J. Reine Angew. Math.* **311/312** (1979), 145-169.

[39] R. Hempel & J. Voigt. The spectrum of a Schrödinger operator in $L_p(R^\nu)$ is p-independent, *Comm. Math. Phys.* **104** no. 2 (1986), 243–250.

[40] M. Hieber. Gaussian estimates and holomorphy of semigroups on L^p spaces, *J. London Math. Soc.* (2) **54** no. 1 (1996), 148–160.

[41] S. Hofmann & A. McIntosh. The solution of the Kato problem in two dimensions, Proceedings of the conference on harmonic analysis and PDE held at El Escurial, June 2000, Pub. Math. Vol. extra, 2002, pp 143-160.

[42] S. Hofmann & J.M. Martell. L^p bounds for Riesz transforms and square roots associated to second order elliptic operators, *Publ. Mat.* **47** (2003), 497-515.

[43] L. Hörmander. Estimates for translation invariant operators in L^p spaces, *Acta Math.* **104** (1960), 93-140.

[44] T. Iwaniec & C. Sbordone. Riesz transforms and elliptic PDEs with VMO coefficients, *J. Anal. Math.* **74** (1998), 183–212.

[45] T. Kato. Fractional powers of dissipative operators, *J. Math. Soc. Japan* **13** (1961), 246-274.

[46] T. Kato. *Perturbation theory for linear operators.* Springer Verlag, New York, 1966.

[47] C. Kenig. *Harmonic Analysis techniques for second order elliptic boundary value problems*, volume 83 of *CBMS - conference lecture notes.* AMS, Providence RI, 1995.

[48] C. Le Merdy. On square functions associated to sectorial operators, *Bull. Soc. Math. France* **132** no. 1 (2004), 137–156.

[49] J.-L. Lions. Espaces d'interpolation et domaines de puissances fractionnaires, *J. Math. Soc. Japan* **14** (1962), 233–241.

[50] V. Liskevitch, Z. Sobol & H. Vogt. On L_p-theory of C_0 semigroups associated with second order operators. II, *J. Funct. Anal.* **193** (2002) 55–76.

[51] J. M. Martell. Sharp maximal functions associated with approximations of the identity in spaces of homogeneous type and applications, *Studia Math.* **161** (2004), 113–145.

[52] V.G. Maz'ya, S.A. Nazarov & B.A. Plamenevskii. Absence of De Giorgi-type theorems for strongly elliptic equations with complex coefficients, *J. Math. Sov.* **28** (1985), 726–739.

[53] A. McIntosh. Operators which have an H^∞ functional calculus. In *Miniconference on operator theory and partial differential equations*, volume 14 of *Center for Math. and Appl.*, pages 210–231, Canberra, 1986. Australian National Univ.

[54] A. McIntosh & A. Yagi. Operators of type ω without a bounded H^∞ functional calculus, In *Miniconference on Operators in Analysis*, volume 24 of *Center for Math. and Appl.*, pages 159–174, Canberra, 1989. Australian National Univ.

[55] Y. Meyer. *Ondelettes et opérateurs*. Hermann, volumes 1 & 2, 1990.

[56] Y. Meyer & R. Coifman. *Ondelettes et opérateurs*. Hermann, volume 3, 1990.

[57] C. Morrey. *Multiple integrals in the calculus of variations*. Springer Verlag, 1966.

[58] El-Maati Ouhabaz. Gaussian estimates and holomorphy of semigroups, *Proc. Amer. Math. Soc.* **123** no. 5 (1995), 1465–1474.

[59] Z. Shen. Bounds of Riesz transforms on L^p spaces for second order elliptic operators, *Ann. Inst. Fourier* **55** no 1 (2005), 173–197.

[60] I. Sneiberg. Spectral properties of linear operators in interpolation families of Banach spaces, *Mat. Issled.* **9** (1974), 214–229.

[61] G. Stampacchia. The spaces $\mathcal{L}^{(p,\lambda)}$, $N^{(p,\lambda)}$ and interpolation, *Ann. Scuola Norm. Sup. Pisa* (3) **19** (1965), 443–462.

[62] E.M. Stein. *Singular integrals and differentiability of functions*. Princeton Univ. Press, 1970.

[63] E.M. Stein. *Harmonic Analysis, real-variable methods, orthogonality, and oscillatory integrals*. Princeton Univ. Press, 1993.

[64] E.M. Stein & G. Weiss. *Introduction to Fourier analysis on Euclidean space*. Princeton Univ. Press, 1971.

[65] N.T. Varopoulos, L. Saloff-Coste & T. Coulhon. *Analysis and geometry on groups*, volume 100 of *Cambridge Tracts in Mathematics*. Cambridge University Press, 1993.

[66] J. Verdera. The fall of the doubling condition in harmonic analysis, *Publ. Mat.*, Vol. extra (2002), 275–292.

[67] A. Yagi. Coïncidence entre des espaces d'interpolation et des domaines de puissances fractionnaires d'opérateurs, *C. R. Acad. Sci. Paris*, Série 1, **299** (1984), 173–176.

[68] L. Yan. Littlewood-Paley functions associated to second order operators, *Math. Z.* **246** (2004), 655-666.

Editorial Information

To be published in the *Memoirs*, a paper must be correct, new, nontrivial, and significant. Further, it must be well written and of interest to a substantial number of mathematicians. Piecemeal results, such as an inconclusive step toward an unproved major theorem or a minor variation on a known result, are in general not acceptable for publication.

Papers appearing in *Memoirs* are generally at least 80 and not more than 200 published pages in length. Papers less than 80 or more than 200 published pages require the approval of the Managing Editor of the Transactions/Memoirs Editorial Board.

As of November 30, 2006, the backlog for this journal was approximately 12 volumes. This estimate is the result of dividing the number of manuscripts for this journal in the Providence office that have not yet gone to the printer on the above date by the average number of monographs per volume over the previous twelve months, reduced by the number of volumes published in four months (the time necessary for preparing a volume for the printer). (There are 6 volumes per year, each usually containing at least 4 numbers.)

A Consent to Publish and Copyright Agreement is required before a paper will be published in the *Memoirs*. After a paper is accepted for publication, the Providence office will send a Consent to Publish and Copyright Agreement to all authors of the paper. By submitting a paper to the *Memoirs*, authors certify that the results have not been submitted to nor are they under consideration for publication by another journal, conference proceedings, or similar publication.

Information for Authors

Memoirs are printed from camera copy fully prepared by the author. This means that the finished book will look exactly like the copy submitted.

Initial submission. The AMS uses Centralized Manuscript Processing for initial submissions. Authors should submit a PDF file using the Initial Manuscript Submission form found at **www.ams.org/cgi-bin/peertrack/submission.pl**, or send one copy of the manuscript to the following address: Centralized Manuscript Processing, MEMOIRS OF THE AMS, 201 Charles Street, Providence, RI 02904-2294 USA. If a paper copy is being forwarded to the AMS, indicate that it is for it Memoirs and include the name of the corresponding author, contact information such as email address or mailing address, and the name of an appropriate Editor to review the paper (see the list of Editors below).

The paper must contain a *descriptive title* and an *abstract* that summarizes the article in language suitable for workers in the general field (algebra, analysis, etc.). The *descriptive title* should be short, but informative; useless or vague phrases such as "some remarks about" or "concerning" should be avoided. The *abstract* should be at least one complete sentence, and at most 300 words. Included with the footnotes to the paper should be the 2000 *Mathematics Subject Classification* representing the primary and secondary subjects of the article. The classifications are accessible from **www.ams.org/msc/**. The list of classifications is also available in print starting with the 1999 annual index of *Mathematical Reviews*. The Mathematics Subject Classification footnote may be followed by a list of *key words and phrases* describing the subject matter of the article and taken from it. Journal abbreviations used in bibliographies are listed in the latest *Mathematical Reviews* annual index. The series abbreviations are also accessible from **www.ams.org/publications/**. To help in preparing and verifying references, the AMS offers MR Lookup, a Reference Tool for Linking, at **www.ams.org/mrlookup/**.

Electronically prepared manuscripts. The AMS encourages electronically prepared manuscripts, with a strong preference for $\mathcal{A}_{\mathcal{M}}\mathcal{S}$-LaTeX. To this end, the Society has prepared $\mathcal{A}_{\mathcal{M}}\mathcal{S}$-LaTeX author packages for each AMS publication. Author packages include instructions for preparing electronic manuscripts, samples, and a style file that generates the particular design specifications of that publication series. Though $\mathcal{AMS}$-LATEX is the highly preferred format of TEX, author packages are also available in $\mathcal{AMS}$-TEX.

Authors may retrieve an author package from the AMS website starting from www.ams.org/tex/ or via FTP to ftp.ams.org (login as anonymous, enter username as password, and type cd pub/author-info). The *AMS Author Handbook* and the *Instruction Manual* are available in PDF format following the author packages link from www.ams.org/tex/. The author package can also be obtained free of charge by sending email to tech-support@ams.org (Internet) or from the Publication Division, American Mathematical Society, 201 Charles St., Providence, RI 02904-2294, USA. When requesting an author package, please specify $\mathcal{AMS}$-LATEX or $\mathcal{AMS}$-TEX and the publication in which your paper will appear. Please be sure to include your complete mailing address.

After acceptance. The final version of the electronic file should be sent to the Providence office (this includes any TEX source file, any graphics files, and the DVI or PostScript file) immediately after the paper has been accepted for publication.

Before sending the source file, be sure you have proofread your paper carefully. The files you send must be the EXACT files used to generate the proof copy that was accepted for publication. For all publications, authors are required to send a printed copy of their paper, which exactly matches the copy approved for publication, along with any graphics that will appear in the paper.

Accepted electronically prepared files can be submitted via the web at www.ams.org/submit-book-journal/, sent via FTP, or sent on CD-Rom or diskette to the Electronic Prepress Department, American Mathematical Society, 201 Charles Street, Providence, RI 02904-2294 USA. TEX source files, DVI files, and PostScript files can be transferred over the Internet by FTP to the Internet node ftp.ams.org (130.44.1.100). When sending a manuscript electronically via CD-Rom or diskette, please be sure to include a message identifying the paper as a Memoir.

Electronically prepared manuscripts can also be sent via email to pub-submit@ams.org (Internet). In order to send files via email, they must be encoded properly. (DVI files are binary and PostScript files tend to be very large.)

Electronic graphics. Comprehensive instructions on preparing graphics are available at www.ams.org/jourhtml/. A few of the major requirements are given here.

Submit files for graphics as EPS (Encapsulated PostScript) files. This includes graphics originated via a graphics application as well as scanned photographs or other computer-generated images. If this is not possible, TIFF files are acceptable as long as they can be opened in Adobe Photoshop or Illustrator. No matter what method was used to produce the graphic, it is necessary to provide a paper copy to the AMS.

Authors using graphics packages for the creation of electronic art should also avoid the use of any lines thinner than 0.5 points in width. Many graphics packages allow the user to specify a "hairline" for a very thin line. Hairlines often look acceptable when proofed on a typical laser printer. However, when produced on a high-resolution laser imagesetter, hairlines become nearly invisible and will be lost entirely in the final printing process.

Screens should be set to values between 15% and 85%. Screens which fall outside of this range are too light or too dark to print correctly. Variations of screens within a graphic should be no less than 10%.

Inquiries. Any inquiries concerning a paper that has been accepted for publication should be sent to memo-query@ams.org or directly to the Electronic Prepress Department, American Mathematical Society, 201 Charles St., Providence, RI 02904-2294 USA.

Editors

This journal is designed particularly for long research papers, normally at least 80 pages in length, and groups of cognate papers in pure and applied mathematics. Papers intended for publication in the *Memoirs* should be addressed to one of the following editors. The AMS uses Centralized Manuscript Processing for initial submissions to AMS journals. Authors should follow instructions listed on the Initial Submission page found at www.ams.org/memo/memosubmit.html.

Algebra to ALEXANDER KLESHCHEV, Department of Mathematics, University of Oregon, Eugene, OR 97403-1222; email: ams@noether.uoregon.edu

Algebra and its application to MINA TEICHER, Emmy Noether Research Institute for Mathematics, Bar-Ilan University, Ramat-Gan 52900, Israel; email: teicher@macs.biu.ac.il

Algebraic geometry to DAN ABRAMOVICH, Department of Mathematics, Brown University, Box 1917, Providence, RI 02912; email: amsedit@math.brown.edu

Algebraic number theory to V. KUMAR MURTY, Department of Mathematics, University of Toronto, 100 St. George Street, Toronto, ON M5S 1A1, Canada; email: murty@math.toronto.edu

Algebraic topology to ALEJANDRO ADEM, Department of Mathematics, University of British Columbia, Room 121, 1984 Mathematics Road, Vancouver, British Columbia, Canada V6T 1Z2; email: adem@math.ubc.ca

Combinatorics to JOHN R. STEMBRIDGE, Department of Mathematics, University of Michigan, Ann Arbor, Michigan 48109-1109; email: FRS@umich.edu

Complex analysis and harmonic analysis to ALEXANDER NAGEL, Department of Mathematics, University of Wisconsin, 480 Lincoln Drive, Madison, WI 53706-1313; email: nagel@math.wisc.edu

Differential geometry and global analysis to LISA C. JEFFREY, Department of Mathematics, University of Toronto, 100 St. George St., Toronto, ON Canada M5S 3G3; email: jeffrey@math.toronto.edu

Dynamical systems and ergodic theory to AMIE WILKINSON, Department of Mathematics, Northwestern University, 2033 Sheridan Road, Evanston, IL 60208-2730; email: transactions@math.northwestern.edu

Functional analysis and operator algebras to DIMITRI SHLYAKHTENKO, Department of Mathematics, University of California, Los Angeles, CA 90095; email: shlyakht@math.ucla.edu

Geometric analysis to WILLIAM P. MINICOZZI II, Department of Mathematics, Johns Hopkins University, 3400 N. Charles St., Baltimore, MD 21218; email: trans@math.jhu.edu

Geometric analysis to MLADEN BESTVINA, Department of Mathematics, University of Utah, 155 South 1400 East, JWB 233, Salt Lake City, Utah 84112-0090; email: bestvina@math.utah.edu

Harmonic analysis, representation theory, and Lie theory to ROBERT J. STANTON, Department of Mathematics, The Ohio State University, 231 West 18th Avenue, Columbus, OH 43210-1174; email: stanton@math.ohio-state.edu

Logic to STEFFEN LEMPP, Department of Mathematics, University of Wisconsin, 480 Lincoln Drive, Madison, Wisconsin 53706-1388; email: lempp@math.wisc.edu

Partial differential equations to GUSTAVO PONCE, Department of Mathematics, South Hall, Room 6607, University of California, Santa Barbara, CA 93106; email: ponce@math.ucsb.edu

Partial differential equations and dynamical systems to PETER POLACIK, School of Mathematics, University of Minnesota, Minneapolis, MN 55455; email: polacik@math.umn.edu

Probability and statistics to KRZYSZTOF BURDZY, Department of Mathematics, University of Washington, Box 354350, Seattle, Washington 98195-4350; email: burdzy@math.washington.edu

Real analysis and partial differential equations to DANIEL TATARU, Department of Mathematics, University of California, Berkeley, Berkeley, CA 94720; email: tataru@math.berkeley.edu

All other communications to the editors should be addressed to the Managing Editor, ROBERT GURALNICK, Department of Mathematics, University of Southern California, Los Angeles, CA 90089-1113; email: guralnic@math.usc.edu.

Titles in This Series

875 **C. Krattenthaler and T. Rivoal,** Hypergéométrie et fonction zêta de Riemann, 2007

874 **Sonia Natale,** Semisolvability of semisimple Hopf algebras of low dimension, 2007

873 **A. J. Duncan,** Exponential genus problems in one-relator products of groups, 2007

872 **Anthony V. Geramita, Tadahito Harima, Juan C. Migliore, and Yong Su Shin,** The Hilbert function of a level algebra, 2007

871 **Pascal Auscher,** On necessary and sufficient conditions for L^p-estimates of Riesz transforms associated to elliptic operators on $\mathbb{R}^n$ and related estimates, 2007

870 **Takuro Mochizuki,** Asymptotic behaviour of tame harmonic bundles and an application to pure twistor D-modules, Part 2, 2007

869 **Takuro Mochizuki,** Asymptotic behaviour of tame harmonic bundles and an application to pure twistor D-modules, Part 1, 2007

868 **Gelu Popescu,** Entropy and multivariable interpolation, 2006

867 **Vilmos Totik,** Metric properties of harmonic measures, 2006

866 **William Craig,** Semigroups underlying first-order logic, 2006

865 **Nathanial P. Brown,** Invariant means and finite representation theory of $C*$-algebras, 2006

864 **John M. Lee,** Fredholm operators and Einstein metrics on conformally compact manifolds, 2006

863 **M. Lübke and A. Teleman,** The Universal Kobayashi-Hitchin correspondence on Hermitian manifolds, 2006

862 **Alberto Canonaco,** The Beilinson complex and canonical rings of irregular surfaces, 2006

861 **Leon A. Takhtajan and Lee-Peng Teo,** Weil-Petersson metric on the universal Teichmüller space, 2006

860 **Thomas M. Fiore,** Pseudo limits, biadjoints and pseudo algebras: Categorical foundations of conformal field theory, 2006

859 **N. Arcozzi, R. Rochberg, and E. Sawyer,** Carleson measures and interpolating sequences for Besov spaces on complex balls, 2006

858 **Enrico Valdinoci, Berardino Sciunzi, and Vasile Ovidiu Savin,** Flat level set regularity of p-Laplace phase transitions, 2006

857 **Donatella Danielli, Nocola Garofalo, and Duy-Minh Nhieu,** Non-doubling Ahlfors measures, perimeter measures, and the characterization of the trace spaces of Sobolev functions in Carnot-Carathéodory spaces, 2006

856 **Vladimir Bolotnikov and Harry Dym,** On boundary interpolation for matrix valued Schur functions, 2006

855 **Yevgenia Kashina, Yorck Sommerhäuser, and Yongchang Zhu,** On higher Frobenius-Schur indicators, 2006

854 **Noam Greenberg,** The role of true finiteness in the admissible recursively enumerable degrees, 2006

853 **Joachim Krieger,** Stability of spherically symmetric wave maps, 2006

852 **Viorel Barbu, Irena Lasiecka, and Roberto Triggiani,** Tangential boundary stabilization of Navier-Stokes equations, 2006

851 **Jie Wu,** On maps from loop suspensions to loop spaces and the shuffle relations on the Cohen groups, 2006

850 **Siegfried Echterhoff, S. Kaliszewski, John Quigg, and Iain Raeburn,** A categorical approach to imprimitivity theorems for C^*-dynamical systems, 2006

849 **Katsuhiko Kuribayashi, Mamoru Mimura, and Tetsu Nishimoto,** Twisted tensor products related to the cohomology of the classifying spaces of loop groups, 2006

848 **Bob Oliver,** Equivalences of classifying spaces completed at the prime two, 2006

TITLES IN THIS SERIES

847 **Eric T. Sawyer and Richard L. Wheeden,** Hölder continuity of weak solutions to subelliptic equations with rough coefficients, 2006

846 **Victor Beresnevich, Detta Dickinson, and Sanju Velani,** Measure theoretic laws for lim–sup sets, 2006

845 **Ehud Friedgut, Vojtech Rödl, Andrzej Ruciński, and Prasad V. Tetali,** A Sharp threshold for random graphs with a monochromatic triangle in every edge coloring, 2006

844 **Amadeu Delshams, Rafael de la Llave, and Tere M. Seara,** A geometric mechanism for diffusion in Hamiltonian systems overcoming the large gap problem: Heuristics and rigorous verification on a model, 2006

843 **Denis V. Osin,** Relatively hyperbolic groups: Intrinsic geometry, algebraic properties, and algorithmic problems, 2006

842 **David P. Blecher and Vrej Zarikian,** The calculus of one-sided M-ideals and multipliers in operator spaces, 2006

841 **Enrique Artal Bartolo, Pierrette Cassou-Noguès, Ignacio Luengo, and Alejandro Melle Hernández,** Quasi-ordinary power series and their zeta functions, 2005

840 **Sławomir Kołodziej,** The complex Monge-Ampère equation and pluripotential theory, 2005

839 **Mihai Ciucu,** A random tiling model for two dimensional electrostatics, 2005

838 **V. Jurdjevic,** Integrable Hamiltonian systems on complex Lie groups, 2005

837 **Joseph A. Ball and Victor Vinnikov,** Lax-Phillips scattering and conservative linear systems: A Cuntz-algebra multidimensional setting, 2005

836 **H. G. Dales and A. T.-M. Lau,** The second duals of Beurling algbras, 2005

835 **Kiyoshi Igusa,** Higher complex torsion and the framing principle, 2005

834 **Keníchi Ohshika,** Kleinian groups which are limits of geometrically finite groups, 2005

833 **Greg Hjorth and Alexander S. Kechris,** Rigidity theorems for actions of product groups and countable Borel equivalence relations, 2005

832 **Lee Klingler and Lawrence S. Levy,** Representation type of commutative Noetherian rings III: Global wildness and tameness, 2005

831 **K. R. Goodearl and F. Wehrung,** The complete dimension theory of partially ordered systems with equivalence and orthogonality, 2005

830 **Jason Fulman, Peter M. Neumann, and Cheryl E. Praeger,** A generating function approach to the enumeration of matrices in classical groups over finite fields, 2005

829 **S. G. Bobkov and B. Zegarlinski,** Entropy bounds and isoperimetry, 2005

828 **Joel Berman and Paweł M. Idziak,** Generative complexity in algebra, 2005

827 **Trevor A. Welsh,** Fermionic expressions for minimal model Virasoro characters, 2005

826 **Guy Métivier and Kevin Zumbrun,** Large viscous boundary layers for noncharacteristic nonlinear hyperbolic problems, 2005

825 **Yaozhong Hu,** Integral transformations and anticipative calculus for fractional Brownian motions, 2005

824 **Luen-Chau Li and Serge Parmentier,** On dynamical Poisson groupoids I, 2005

823 **Claus Mokler,** An analogue of a reductive algebraic monoid whose unit group is a Kac-Moody group, 2005

822 **Stefano Pigola, Marco Rigoli, and Alberto G. Setti,** Maximum principles on Riemannian manifolds and applications, 2005

For a complete list of titles in this series, visit the
AMS Bookstore at **www.ams.org/bookstore/**.